FOSSIL VERTEBRATES -

Beach and Bank Collecting for Amateurs

By M. C. Thomas
Author of "LET'S FIND FOSSILS ON THE BEACH"

For T., for everything.

Sixteenth Printing March, 1981

ACKNOWLEDGMENTS:

I wish to express my appreciation to:

Dr. S. David Webb
Mr. John S. Waldrop
Mr. Norman T. Tessman
Department of Natural Sciences
Florida State Museum
Gainesville, Florida

for their encouragement and concern in this work and for their aid in determining the identification of some of the specimens photographed for this book.

Naturally, they are not responsible for any errors that may have crept into this effort.

Grateful acknowledgment is extended to Mr. Howard H. Converse, Jr., for the photography and for his additional help.

Fossil material for pictures of specimens, other than those in the author's collection, were loaned by amateurs and professionals who are listed alphabetically, with many thanks:

Mr. Albert Barney
Mr. and Mrs. Roy A. Burgess
Mr. Howard H. Converse, Jr.
Mr. and Mrs. William R. Hall
Mrs. Malcolm Taylor
Mr. Norman T. Tessman
Mr. John S. Waldrop

Deep appreciation is extended to all the paleontologists and geologists and amateurs who have provided the data concerning their state's coastal beaches and banks.

Continuous gratitude goes to the Ph.D. who has helped so much so long. At his request, and because of his position, he must remain anonymous.

PREFACE

This is a primer and guide for amateur beach and bank combers of all ages who would like to collect fossil vertebrates but who don't know:

WHERE

WHEN

HOW

or

WHY

to begin

Take a pail to a fossiliferous beach or bank and you can fill it with diversified specimens of the remains of the vertebrate animal world of historic and pre-historic times. Some beaches are listed at the end of the book. Start at the beginning with the BEACH hunting section and when you finish that, you will be ready to progress to Part II . . . AND BANK.

FROM ONE AMATEUR COLLECTOR TO ANOTHER

"The Pleistocene history of North America holds ten major mysteries for every one that has already been solved."

R. A. Daly
Harvard University

Many of the mysteries can be explained by knowledge of normal geologic and biologic processes. The solution to others is still being debated by scientists. For instance, some say that the extinction of many of the animals was caused by world-wide catastrophes and upheavals. Others say that this is a theory, not a fact, and that man (then on the scene) must shoulder much of the blame.

There are many issues concerning the epoch of the Ice Ages that remain in contention and the research and study continues.

This is a primer for the hobby of hunting the remains of ancient vertebrate animals on beaches and banks around the perimeter of the United States. It answers no scientific questions. However, it is written with the hope that your interest will be stimulated in the earth sciences so that you will want to dig deeper into the past. Perhaps it will be you who finds a key to one of the doors behind which lie some of the answers.

Curiosity is the first step to any discovery. Wondering must have a beginning, so stop and read and begin to wonder. It's fun.

M. C. THOMAS

PART 1

BEACH . . .

CHAPTER I

There were four major glacial advances during the Ice Ages. These Ice Ages occurred during the Pleistocene epoch (which is believed to have begun one to three million years ago) and which is part of the Cenozoic era. During these advances existing glaciers and ice caps (such as those on Greenland and Antarctica) expanded and new ones formed, and snow lines were lowered over much of the world. There were numerous centers of glaciation which pushed into low latitudes and low altitudes when the weather was colder and retreated into higher latitudes and higher altitudes when the weather was warmer. During these warm phases the weather was, at times, warmer than it is today. Although the ice sheets disappeared (from what is now the United States) 10,000 years ago, many geologists feel that we are living in the Pleistocene still and that our time is actually another interglacial period.

During the freezes continental ice sheets came as far south as the Ohio and the Missouri Rivers in the eastern United States. The western beaches were free of ice. So were many of the ones on the east coast.

Some animals that died in the Arctic were occasionally preserved, including their fur and flesh, in frozen river gravel and mud. These, far too few specimens, are hunted avidly by the scientist and upon discovery are most important for study to aid in the furtherance of knowledge of the past.

There was, at the beginning of the Ice Ages, a land bridge which joined Alaska and Siberia. It is now the Bering Strait. Over this, before and during the Ice Ages, animals crossed. The eastward bound joined the native beasts of North America. As time and generations went by some of these beasts moved south. The elephants, various species of cats, the rhinoceroses and other ancient beasts and birds were already part of the scene. In fact the rhinos, which may have originated in North America, had been here and were gone from North America before the Pleistocene. This does not mean that you cannot find their bones and teeth occasionally but you are mainly interested in those animals that were near your coasts during the Ice Ages—the Pleistocene epoch. Pre-Pleistocene finds will be the frosting on the cake which is truly very rich in itself.

As does the present day tourist, many of the vertebrates (animals that have backbones or spinal columns) came to the shore if they happened to live close enough to it; some lived there all their lives and some died there. It is their fossilized remains that you can find.

When the northlands became frozen the waters of the U. S. coastlines receded from the shores. They accumulated in the form of snow in the north. During these times coastal land that previously had been under water was exposed. During the thawing periods many beaches of today were under water. Each of these conditions existed for different periods of time depending on the weather in the north and on the rise and fall of land that still takes place.

In this guide to finding fossils on beaches you will be interested in the periods during which dry or marshy land extended from feet to miles into what is now ocean, gulf, bay or river mouth. Though plants and life continued in the far north, certainly a more temperate climate was enjoyed farther south. Here a greater variety of foods could grow and here beasts could exist in relative comfort.

Not in your lifetime will you ever see such a conglomeration and variety of living animals outside of a large zoo.

Some animals became the ancestors of the birds and beasts of today. Some became extinct because of unusual food or living conditions to which they could not adjust. Still others disappeared from the American scene for reasons that are still a mystery to man. A case in point is the ancestor of the present day horse, which is probably included in the same species. He is known as *Equus*.

You were taught that the Spaniards brought the horse to America. This is true but it was a return trip for the horse. Until about 10,000 years ago horses had lived here for millions of years and had developed from tiny primitive species to forms essentially identical to the modern horse. Then he became extinct in the New World — why, nobody knows. He continued to live and thrive in the Old World and it was his many times great, great, grand colts and fillies that became the transportation for the Spaniards after they anchored their ships in harbors that now belong to the United States.

CHAPTER II

During the Pleistocene epoch the larger animals were not alone. There were the ancestors of all the animals that live today. These include the modern bears, wolves, dogs, cats (huge saber-toothed cats down through the pumas to the size of a modern-day cat), peccaries, deer, horses, camels, guanacos and tapirs to mention but a few. Rabbits, mice, giant beavers, squirrels, raccoons, bison, goats, snakes, birds, fishes, and you-name-it. When you hear the expression "Age of Mammals" it refers to the Cenozoic era of which you are particularly concerned with the Pleistocene epoch.

As generations went by, some of the land species emigrated to South America. Others remained in the United States and some animals went farther north when the warmer weather came. The record of all of this can be found in fossil form.

The immigration to South America and the emigration from South America to the North American continent were made possible by another land bridge. South America had been separated from Central America for seventy million years — it had been an island during almost all of the "Age of Mammals". At the beginning of the Ice Ages the Isthmus of Panama had risen from the sea. This became the bridge. It was not only popular but must have been fairly crowded. The animal species native to South America were many. Some of their porcupines, opossums, ground sloths, armadillos, spider monkeys, capybaras (a large rodent which is now extinct in North America), and glyptodonts (huge relatives of the armadillo, now extinct everywhere) were among those to start north. On their way, they met the beasts from the north that were eventually going to make their homes in South America. Many habitats were traded. Some dogs, pumas, deer, tapirs, raccoons, horses, saber-toothed cats, guanacos, squirrels, spiny rats and some of the bears went to South America. Some stayed home. Some left us for good. Some never went back to South America — others did. Some animals, among which were the giant anteaters, agoutis, toxodonts and two-toed sloths started up from South America, got as far as Central America and settled there. You do not find their remains on the beaches. It was a busy time. It took many generations for each animal to move to a new location. No animal covered thousands of miles in his own lifetime.

The mastodon was elephantine in appearance but not as closely related to the modern elephants as was the mammoth. When you see the difference between the teeth, alone, of the mastodon and the present day elephants, and the similarity between those of the mammoth and the modern elephants you will be able to understand why science has given the nod to the mammoth. There were many developments and changes in each of these beasts before the Pleistocene and several species of each lived during the Ice Ages. You may call them simply mammoths and mastodons and their various bones, teeth and tusks, "ELEPHANT MATERIAL". It is most difficult, for an amateur collector, to decide what fossil specimen (other than teeth) comes from which animal unless you happen to have the skeleton of one of each on hand for purposes of comparison. Use your museums and learn from what you can see in them.

Both the mammoths and the mastodons lived, during the Pleistocene, in what has now become the United States. Almost certainly only the mastodons were ever in South America. The mammoths stayed here and in the far north in the New World.

Many animals died of old age, improper diet or injuries. Some died through carelessness and some as a result of hunger, or avarice. You can easily imagine some huge creature, perhaps a mammoth, becoming bogged in marshy ground. During his thrashings and trumpetings he attracted the attention of a hungry saber-toothed cat. The cat could have been the size of a modern tiger and very unlike his domesticated descendants. He had long pointed and curved fangs that protruded from his upper jaw. The jawbone structure enabled him to drop his lower jaw to wider than a right angle from his upper. Then the fangs were cleared for action and became excellent weapons for puncturing and tearing his prey. Long had this cat waited for such a huge and safely immobolized meal. He sprang, and the elephant moved. The cat slipped and he, too, was caught in the mud and clay trap. A dire wolf (now extinct), that had been skulking at the edge of the marsh, then added his weight to that of the cat. He, too, began to sink. Smaller and smaller carnivorous (meat eating) beasts attacked the larger creatures and met the same fate. At last, a buzzard joined the pile, at the top, and over a period of time they all became still and the swamp once again, was a quiet place .

You must hunt inland to find closely grouped bones, or those still in the matrix in which they had been buried. To find cleanly separated ones you need only look upon the surface of the beach. The sea has done the digging, the separating and the distributing for your convenience.

Whether the animals died in the marshes or on the shores they lay and then were buried in sand or mud or clay and were protected there, much later to be found as fossils. Because of the actions of the sea you do not find whole fossilized skeletons on the beach. They are broken from their deposits and unearthed little by little.

What is a fossil? A fossil is any evidence of the former existence of an ancient organism. Fossil evidence of vertebrates may consist of bones or teeth but also may consist of a footprint, brain cast, copralite (see end of Chapter III), burrow (worm, rodent, lung-fish, tortoise or clam, etc.), woodrat nest, beaver dam, etc. Mineralization or petrification is irrelevant for your purposes — that means that it doesn't really matter, until you dig deeper into paleontology, whether the fossils that you find have been impregnated with the minerals in which they lay or replaced by them. Some degree of age is always implied, but as we approach the present, the limitations between what is and is not called a fossil become hazy. This is natural as geologic time is continuous. Certainly vertebrate fossils are thousands of years old by the time they can qualify for the term.

The often asked question as to the length of time it takes a bone to become a fossil is impossible to answer without many other questions being answered first. Some of these are: In what kind of material was it originally buried? In what location was it found? How many years was it on land and how many years in the water? How long did it take the deposit to surround it? Don't ask a scientist to give you an answer to the original question unless you and he are prepared to answer many others. Radiocarbon testing is not applicable to beach fossils, but even if it were it would be able to tell you the approximate age of the bone if the animal was no older than 35 thousand years, but not how long it had taken it to become a fossil.

You have now found fossilized bones and it is interesting to note that they appear in a number of shades and colors — black, gray, brown, tan, reddish, cream-colored or off white. Modern bone is white. Sometimes, because of surrounding vegetation, you will find bluish fossilized bones, too. These colors are a result of minerals and vegetation in the deposit in which your find has been held. THE COLORS DO NOT DENOTE AGE NOR IS ONE COLOR MORE "VALUABLE" THAN ANOTHER.

It is fortunate when bones wash ashore if they are in good shape. They have had a long, hard row to hoe.

How old are these bones?

This is a good question. When you find any remains of the rhinoceros or of a three-toed horse, among many others, you will know that they are older than Pleistocene epoch animals as they lived before them. Horse, bison, deer, etc., remains can be thousands of years old or hundreds of thousands or over a million. When you become really interested in this hobby you will be able to dig deep into far more scientific books and learn how to judge ages more accurately. This is the primer for your fun and for you to begin to learn about vertebrate paleontology. This can give you only a taste and a shadow of what has been learned by paleontologists whose life work it is to study the ancient world of animals, their lives, habits, descendants and ancestors. Remember that it is believed by many scientists that the Pleistocene began between one and three million years ago. Some of your finds could be that old. Some animals became extinct as late as eight thousand years ago, so their bones can be that young.

Reputable museums have a scientific habit that you will get used to if you send them bones for identification. The specimens that you send may appear to you in good enough shape to be able to be identified by anyone who is a trained paleontologist. You will at first be distressed, then resigned (but seldom grateful) to see so many of your well wrapped fossils returned to you with the words "possibly", "probably" and "apparently" modifying their identifications. You will be upset, naturally, by their pet word for many of the bones sent to them by amateur collectors — "indeterminate"; "unidentifiable" runs a close second. Through these disappointments you will learn what can be and what cannot be identified. To look on the bright side, some day you may get a note from them concerning a fossil that you have needed help with because you couldn't identify it yourself. This note will tell you what it is with no quibbling and invite you to give it to the museum as a gift! The bone that has meant no more to you than any other up until that time will assume an aura of importance, and you, its finder, will have a feeling of satisfaction and pleasure that certainly helps to cancel out the depression caused by the "duds" that you had been certain were spectacular finds — but weren't! You, the amateur, now have an opportunity to help the pro and his museum or university. Many may learn much from your gift.

CHAPTER III

Camel bones and teeth, too, can be found on some of our fossil beaches. Some of these "Ships of the Desert" originated in North America. The greatest evidence of them is found in the southwestern part of the United States. It is as yet a mystery as to why they disappeared. Far later they were reintroduced by man and they lived well in the deserts. Camels' teeth and bones are harder to find than some of the others but this makes it more worthwhile when you are successful. Most specimens will come from a llama-like camel that was smaller than the larger camels.

A giant ground sloth, the *Eremotherium*, and some smaller relations, the *Megalonyx* and the *Paramylodon*, among others, also wandered near the shores. The big fellow reached a height of twenty feet when he stood up to pull the tender leaves from the tops of the trees. These were a staple food of sloths and their teeth are strong columns that would resemble pieces of almost oval ribs were it not for the delicate lines which encircle their length. The masticating surface is merely a plain (uncomplicated by pattern) indentation. Sloths' bones, teeth and claws are excellent finds. These sloths are extinct, and when they lived there were not as many of them as there were horses, sharks, etc., therefore, their fossilized remains are far more scarce, and harder to find.

Bison bones and teeth are more numerous. The bison of the Pleistocene were, for the most part, plains dwellers. What they were doing near the sea is not known. But they were there and their remains prove it. No bison ever got as far south as South America but traces of him have been found in El Salvador.

Some of these old beasts were very big and horn cores have been found that show a span of six feet from one tip to the other of his horns. The core is the bone inside the horn. It is attached to the skull. The horn is made of soft material and does not last long enough to become a fossil.

The most recent ancestor of the present horse was called the *Equus*. This is nothing more nor less than "horse" in Latin. Here, he will be called "horse", which indeed he was. These horses roamed the forests and plains and were very numerous in the New World. He averaged about the size of an Arab horse of today and seldom reached the size of the modern Percherons. Remains of this latter larger species are more rare as was the animal. However, *Equus* has come a long way beyond his millions of years of ancestors. The earliest horse is thought to have stood about 10 inches high at the shoulder and to have weighed eight or nine pounds. When you have a moment, weigh and measure a typical cat (if there is such a thing) and you'll see that the horse is a healthy and developing breed of animal. He's been around a long time, and you can be glad that you won't be seeing him a few million years from now if he continues at this rate of growth. He doesn't do it fast, though, as that smallest horse lived fifty million years ago — so don't panic.

The *Equus* is notable mainly for its size and for what it did not have — three toes, which were so common in so many of his ancestors. *Equus* had a hoof, such as do the present horses, and the core of this, too, with his teeth and bones, can be found on beaches. The three-toed horses were older than *Equus*. Generally, they were smaller. Their teeth and bones were correspondingly smaller and are quite difficult to find on beaches or inland, though it can be done. There is a common erroneous idea that because an animal is older it must be buried deeper. This is not necessarily so. Earth deposits of different ages vary in depth. Sometimes very old ones are near the surface. It is harder to find the bones and teeth of earlier animals than it is of later animals, because they have been dead longer. They have had more chances to be completely destroyed by time, the elements, and man made and operated machinery.

You do find "little horse" (which covers many families such as *Hipparion* and *Nannippus*) teeth, on the beach though these are from the later species of three-toed horses. It is particularly interesting to note that the masticating, or biting and chewing, surface patterns of *Nannippus* teeth are far less ornate and complicated than that which nature believes is necessary and practical for *Equus* and for today's horse.

This leads to a discussion of teeth, without specimens of which so much less would be known by man about ancient animals — including himself.

The size and shape of the tooth certainly helps you to judge whether it came from a large or a small animal. The masticating surface of the tooth gives you the necessary clues as to whether the animal ate meat or was a grazer or both. The meat eater has sharp teeth for killing, tearing and chewing its prey. The grazer does not. The pattern can be understood by the experts and it identifies different species of animals for them. At first, this will all be difficult for you to understand. However, as soon as you have found enough fossils' teeth, you will be able to compare them and you will see the very apparent differences and shortly be able to identify the animals from which they came. Teeth don't break as easily as bones and you can see what varieties of animals were here by just finding their teeth. Many bones have many similarities. Teeth of each different kind of animal have individual patterns that are his alone.

The enamel on the teeth is the original enamel. It is fossilized, but in general, owing to its dense structure, it is not impregnated with (or replaced by) minerals as are the bones.

While you hunt on the beaches, it is interesting to remember that every animal that walks, swims or flies now (or does a combination of these) has an ancestor that lived during the "Ice Ages". Some have ancestors that can be traced back into the dim past millions of years before the Ice Ages. Others, such as the manatee (sea-cow), are real orphans. As yet, very little, if any, is known about his background. He appeared during the Pleistocene.

It is the work of the vertebrate paleontologist, with his study of ancient bones, to unravel the history of vertebrates. (invertebrate paleontologists and paleobotanists do similar work with animals without backbones and with plants, respectively). From the fossilized skeleton he can better understand the live creature. Where it is found helps the earth scientists to judge the climate of the times. It also helps to discover and then to confirm the facts concerning the geography of lands and seas of the world when the animal lived.

Other important signposts of the past are fossilized fecal pellets. These are called coprolites. From them, and by scientifically testing their composition, the paleontologists and biologists can learn a great deal about the feeding habits of the animals. They are also clues to what was growing in those days and, therefore, what the temperature probably approximated.

Paleontology is an integral part of other earth sciences, such as biology and geology. Without the knowledge that has been gained through it, the writings concerning pre-history would still be in outline form.

CHAPTER IV

On the beach you will find more bits and pieces of ribs than anything else. The diameter of some of these is as great as three to four inches. The ribs that are particularly well rounded and heavy and dense (no marrow holes at the breaks) are probably of a sea-cow. If this rib comes from the

Pleistocene epoch, it is from the manatee. If it is pre-Pleistocene, it is from the dugong. Therefore, sea-cow is a good general term. No manatees were heard of previous to the Pleistocene epoch — another mystery — how and from what did they get started? One species of dugong still swims in the Indian Ocean and the Red Sea. The dugong of which you find fossilized parts is extinct.

Ivory from the tusked animals such as elephant and walrus will be thinly spread, but well worth hunting on the beaches. It is usually found in small slivered pieces, though occasionally you will be lucky enough to find a tip or a rounded piece of tusk. Ivory can be identified by its velvet smoothness to the touch; it is not shiny like enamel, and the elephant ivory sometimes has cross-hatching (xxx's) on a broken edge. Only elephant tusk ivory has it. Many tusk fragments from elephants don't.

You will learn the difference in "feel" between bone, enamel, and ivory. Ivory breaks in layers rather than cleanly as do most ribs and bones. In the middle of a piece of tusk is a hole that was originally filled with pulp. Many people think that they have found a piece of petrified wood when they find a piece of tusk. This is easy to understand, for if you have a rounded center section of tusk you can see growth rings on the ends similar to those of a tree. Ivory is far smoother to the touch than any unpolished piece of ancient wood.

Curved pieces of very smooth shiny material are enamel. These pieces are broken away from the biting surface of the mastodons' (and other animals') teeth. These are called "cusps" and are parts of the humps (cusps) on the top of their teeth. The shape of mastodon teeth makes them susceptible to breakage, and they have so many good places to do it that few complete mastodon teeth are found on the beaches.

The mammoth had huge teeth. Some measure six inches across by eighteen inches long. They are tougher and more ruggedly reinforced than mastodon teeth and are usually able to be found in larger pieces, or after a storm, in their entirety. These teeth are easy to identify. They are the only teeth from a large animal that are made up of upright (as they are placed in the mouth) plates of dentine and enamel, each partially covered with cement. The substances of which the mammoth teeth are composed go all the way through the tooth. The enamel is seldom shiny except for that on the masticating surface. Mammoths sometimes had as many as thirty plates (or layers) in each of their back teeth. It is fun to find mammoth molars. It is a great find if you locate a milk tooth of one of these huge beasts. These can be as small as one by one and a half inches. See them in the illustrations. There is no enamel in the composition of these teeth. They are much scarcer and far more difficult to find.

A really old vertebrate is the shark. Sharks' teeth are found in rock of the Devonian period which was more than three hundred million years ago. When you find a shark's tooth that has washed in from the sea, you must realize that it is extremely difficult for an amateur to date it. Some animals' teeth can be dated more easily because of the very distinctive changes that occurred in them from epoch to epoch. The evolution rate of the sharks is very slow. The changes in their teeth are equally slow and demand the inspection of the trained professional eye for an accurate estimation of age.

Nobody can judge age by the condition in which you find a fossil shark's tooth. It can be quite young in the fossil line or extremely old. The actions of the sea can make a fairly young shark's tooth look old and battered. A tooth that has been snugly buried for millions of years may wash in with its point, top, enamel and serrated edges in excellent shape.

You may wonder why you hear little of fossilized shark skeletons that have been found by the paleontologists. No sharks had bones in the early days and they still don't have any. The hard parts of sharks and of their relatives, the sting rays and skates, are composed of cartilage rather than bone. This does not lend itself to preservation of great duration. The largest fossil teeth that have been found are about seven inches long by five inches wide. Once again it has been the teeth that have given science its best picture of the entire animal.

There were and are many different kinds of sharks.

Here are a few of the more common fossil finds:

The largest of all the teeth are from the *Carcharodon megalodon.* He may have reached a length of seventy feet. He is, fortunately, extinct and his only descendant, the white shark, though a man-eater is not as big. See the illustrations for further details. Bear in mind that scientists must estimate the length of the extinct fossil forms. There is no objective basis for an exact determination of their maximum lengths. This would require large, randomly selected, population samples of the fossilized remains which are not available. The estimation of the size of any extinct fossil shark must be based on the size of the largest, isolated fossil teeth. Estimations of the length of ancient sharks that have living descendants is easier. There are quantities of modern shark forms with which to compare the teeth of their ancestors.

Other kinds of ancient sharks were smaller and so were their teeth. These smaller, more delicately formed teeth are easier to find than are the huge ones of the *Carcharodon megalodon.*

Fossil teeth of the ancestors of the modern sharks such as sand, mako, tiger, etc., wash into fossiliferous beaches. There are also many fossil sharks' teeth from the extinct species that have no living descendants.

An average size modern tiger shark may have produced as many as 24,000 teeth during a ten year period. There is no reason to believe that the ancient sharks' teeth that we find on our beaches didn't come from a fish that produced and shed an equal number. Each shark had, and has, hundreds of teeth in his mouth at one time. Only the front row with which he bites are erect. All the others, in several rows, one behind the other, are laid down flat against the interior of the jaws. When an active tooth is lost or shed the tooth in back of it moves into place. This process continues for the life of the shark. It is difficult to believe that the manner in which the teeth of MAN function, in comparison, is a proof to his claim of superiority.

To find the smaller sharks' teeth, look along the edge of the water among the shells and pebbles. These teeth can be found in all sizes from a quarter of an inch to two or more inches long. The very large *Carcharodon* teeth are usually flung high on the beach after a storm. If you are so fortunate as to see one of these washing ashore, hurry to retrieve it. They are so heavy that they will be buried quickly in the sand. It is easier to find the smaller teeth of this huge creature. A two to four inch tooth is a fine one to keep. If you can find one that is five inches long you can guess that it came from a *Carcharodon megalodon* that was about forty feet long.

Sharks' teeth, like the bones and teeth of other fossils of the vertebrate animal world, take on the color of the substance in which they have been buried. They are as white as yours when they are in the live animal. As they become fossils they may turn gray, black, brown, gray-blue, tan, beige or a reddish-brown. You will hear on the beaches that one color or another is older. This is not necessarily so. THE COLOR IS DUE TO THE DEPOSIT IN WHICH THE TEETH HAVE BEEN BURIED. It all depends on the kinds of minerals and vegetation that were in the sea during the time that they were becoming fossils.

There are some fossil sharks' teeth that are as yet unidentified as to a particular species. Compared to the research on other fossil groups, very little shark research has been done. There are few publications that can be understood by amateurs. The scientifically accurate papers are either still in the works or are by pros for pros. The latter who concentrate on fossil shark study are few and far between, also.

The teeth of most of the professionally identified shark species are sufficiently distinctive so that an amateur can determine their origin. However, some of the larger families of sharks contain similar species and their teeth are equally similar. Even a professional may require a large series of teeth to be certain with which particular shark he is dealing. If you can positively recognize and determine these teeth without possessing a whole mouthful of them in a shark's jaw you are no amateur collector. If you can find a modern shark's jaw, study it hard. You will see that each tooth varies to a degree (small or large) from the next one. Many sharks' upper teeth are very different from those in their lower jaws. Front teeth of one kind of shark can be confused with side teeth of another. New teeth can differ from old ones in the same location in either jaw. Different stages of their development in the jaw can also confuse you.

If you are an amateur collector, you may well have to be satisfied to learn to distinguish a tooth as having come from a particular genus of sharks. If you want to know more about it, you will have to send it to a professional for identification.

Because so little professional writing is available to amateurs, there is more amateur guessing and battling in the ranks over sharks' teeth than over any other common fossil find. This is a tremendous waste of time as without the knowledge of what minute (to amateurs) distinctions you are able to see and from which you can determine an identification, you can't win. As the study of sharks' teeth from beasts both extant and extinct increases, more and more knowledge will become available to the amateur. In the meantime be prepared for confusion. It is truly widespread.

The only other part of a shark that you can find in fossil form is the cartilaginous disk from his spinal column. It is hard enough to be preserved and it is usually an inch to two inches across, between a quarter and a half inch thick, circular and resembles an empty bobbin from a sewing machine. The smallest ones may be from a dog fish or some other relatively small relation of a shark.

Here, too, you will be most likely to find parts of the mouth plates of sting rays and skates. These plates are made up of several bars which have deeply serrated surfaces that attach the plate to the mouth. They are used in place of individual teeth. His barbs (or what you want to avoid stepping on in a live one) are there in quantity. You can also find dermal denticles (skin plates) of the sting ray family.

Whales and porpoises, which are closely related to each other, left their vertebrae, teeth, and bones for you to hunt. Whale teeth are a rarity. Porpoise teeth are hard to find but not quite as few and far between as the ivory tooth of a toothed whale. Parts of the inner ear bones of each are on some beaches. Inland, there is more chance of finding the whole thing. However, we call the part we usually find, "an ear".

Whale vertebrae are big and heavy. Porpoise vertebrae are smaller and lighter but look about the same. There is no clear line of demarcation between them for an amateur collector. You have to use your head and your imagination on these. If the vertebra is eight inches tall and five inches wide, or more, it should be obvious to you that a porpoise would sink if he had to carry around a load of them. If it is three inches tall by two inches across, you may be fairly well assured that this wouldn't be much help to a whale.

CHAPTER V

So that you can "check out" a beach quickly to see whether it is fossiliferous, or not, here are a few clues. First, the beach must slope gently into the sea with no sharp drop off, no deep trenches, and no high reefs close to shore. There must be sedimentary rock along the beach. Sedimentary rock is not a solid stone — rather, it is composed of different things that it has gathered unto itself as it is hardening from its original soft stage of sands, limes, marls and other minerals. It is in sedimentary rocks that fossils are buried and from which they break loose. The beach that is perfectly smooth with no sedimentary deposits won't have fossils washing in either. You find some fossils still imbedded in pieces of coral, so be sure to look at any deposits of this, too. Fossiliferous beaches have a good many shells on them. These help the fossils to keep from being buried immediately in the sand. The shells found on a fossiliferous beach are not necessarily fossils.

It is difficult for an amateur collector to understand why the scientific "Powers That Be" in paleontology have not already written many books concerning fossils that may be found on beaches. Books in great number have been written about inland diggings but none include a specific section dealing with beaches. The reason may prove of interest to you. The great majority of fossilized material that is in "museum specimen" condition can (sometimes) be found inland. The professional collector of it is interested in the most perfect examples of each specimen that can be obtained. Why? The answer is easy. It is difficult enough to dig literally into the past without having water-borne material with which to work, to try to date, to attempt to assemble into a skeleton, to display, and from which to teach and to learn. The constant actions of seas, sands, tides and waves, often leave a great deal to be desired in the shape, condition and appearance of a fossilized bone or tooth. Even more important to a scientist is having a history of the fossil enhanced by the earth in which it is found. Beach fossils have none. Their past has been well washed off.

There are times, though, that you do find "museum quality" specimens of individual bones on the beach. You do not find the whole skeleton of any ancient beast lying at your feet on the sand. If you ever think that you do see such a thing, get into the shade in a hurry!

To prove to yourself whether or not a bone is a fossil, take matches or a lighter with you to the beach and you will be all set. If the bone to be tested is wet, let it dry. Then hold a flame under one corner of it for about five seconds. Blow out the fire and smell the bone. If it has no smell, it is a fossil. If it smells like burning hair, it is not.

If you are fortunate you can find teeth in the remains of a jawbone. These, and individual bones in good condition, are fine additions to your collection and to send on as gifts, to museums and to the departments of vertebrate paleontology of your universities and colleges, if you do not want them yourself. Never forget that the paleontologists are most interested in having a refusal, at least, of anything that you consider is a good find. If they don't need it, they can dispose of it or return it to you. Give them a chance at it. However, no professional has any time left over and even throwing away bones takes that. Send only what you have learned is worthwhile and slightly out of the ordinary. You will soon be able to tell what is "good" and what isn't. All pros need innumerable examples of similar fossils to study and to compare. Without them progress grinds to a halt. Your beach collecting can be of great advantage to yourself and to the scientists. What the seas have not already destroyed is yours and for free. Amateurs without any previous study, example or professional guidance, should stay out of the "digs" that are being worked by the pros inland. Amateurs can be a great help to the scientist, on one hand, and they can destroy pre-history in its bed, on the other.

Comparatively little money for expeditions to find ancient bones and teeth is being channeled to the earth sciences, or to the earth scientist. It is going to the sciences dealing with the sky and beyond. Yet it is increasingly important that geologists and biologists continue to gain knowledge of the past for an advanced glimpse of the future. Paleontology is the "timer" for both of these. Without enough specimens, classes in paleontology are like chemistry classes without chemicals or retorts. Neither scientist nor student can test and study or make a collection for display or for comparison purposes, from pictures in a book. What you find and donate may let another student into a vertebrate paleontology class and, if it does, it will start another scientist on his long road of learning so that eventually he will be able to help us all. From your beach finds that you give away, you can start a chain reaction that is pretty impressive. Don't underrate yourself if you go into this hobby of hunting for fossils on beaches!

When, and if, you are unable to identify your finds, write to your nearest State University (Department of Vertebrate Paleontology) or a state or federally supported museum. It is one of the many duties of most of these institutions to aid in the identification (and return to the sender) of such objects as amateurs and professional people send them for this purpose. This is not true of all of such establishments as some may not have the personnel facilities to be able to do it. Therefore, ask before you send. All these people are extremely busy in research, teaching, assembling specimens, exhibits and their field work. However, they are, to a man, very interested in helping amateurs as there is a good chance that you'll be able to turn up, and turn over to them, what proves to be a rarity. If they give you permission to send your fossil bones or teeth, be sure that you wrap them securely. Not only are they brittle with age, but these beach bones have been drying out since you found them and they are not as tough as they appear to be. There will not be any charge for identification but enclose the money for their return in the package. Weigh the box on your scale before you seal it and call your post office to find out the charges. Various departments of our state, federal, and private institutions have budgets and they can spend the money on a lot more worthwhile projects than the return of your material. These things are heavy!

DO NOT rush these people, but if you want your specimens back by a certain date either take them to the "pros", yourself, or ask them in advance how long it will take to complete the identification before you send them. Explain the number, etc., of what you plan to send. Only send bones for identification, or as a gift, that have enough of what you estimate to be, their original shape, apparently original unbroken surfaces, and individuality. Then they can be identified. If they are to be a gift, be sure that you enclose with them the information of where they were found and when. If they are crushed, badly worn, and apparent fragments of what you can perhaps imagine was their original shape and size, not even a genius can identify them. He can say that they are bone, and that is a worthless analysis as you already know that by then, or should.

Be certain that you enclose your name and address inside any package that you send. This will further assure the safe return of your specimens. Insure your package.

To make a collection of your own, or for your local school or town (you'll be surprised to find how many schools and towns don't have them and would appreciate one), will bring you untold satisfaction. You'll get a good tan, stronger arches, and a sense of well being that is seldom found indoors or by just sitting on the beach.

It is all free for the picking except on beaches that are part of state or federally controlled parks or memorials. DON'T hunt on these without personal permission from the ranger or whoever is in charge! There is always a good reason, whether you know it or not, as to why these places are not to have what is found there taken by the public.

Learn early in the game to catalogue your identified bones and teeth properly as to what they are and when and where you found them. Except for the pen, ink and shelves there is little monetary outlay for this hobby. When you begin your collection, keep bits and pieces of what you find until you find more perfect specimens. Don't insist on perfection. It is discouraging to any beginner. When you find excellent specimens be sure to discard (or put in a special pile to give away to interested visitors) the imperfect ones. If you don't, you will soon be swamped in fossils.

CHAPTER VI

Many more animals and parts of animals than are mentioned in this booklet can be found on some fossil bearing beach. Different beaches provide an interesting variety. Usually, it is a part of a bone rather than the whole. Try to find the ends of bones (from which some can be identified) rather than just the middle section. Here, we will only have room to list the things that are the most common of the beach finds, and even some of those, one does not come across every day or even every month. When you find huge bones you can be assured that you probably have pieces of whale or elephant.

It is rather useless to spend time searching the beach at high tide unless there has been a recent storm. That is the only thing that will bring the bones high up on the beach. The sensible approach is to obtain the times of your low tides and go to the beach at dead low tide or certainly within a couple of hours either side of it. The larger bones will almost certainly wait for a heavy sea and a high wind to bring them within your reach if you want to keep your feet dry. They will be washed high onto the beaches and there you can find parts of skulls, leg bones, hip and shoulder bones, elephant and whale vertebrae, elephant teeth and the big shark teeth. Hunt them soon after the storm or during it if you are feeling particularly rugged — after all, the wind that brings them in and the waves that wash them up, will both help to blow sand over them and to bury them in it.

Another way to find larger bones is to wade in the water where there are rocks as well as sand, from ankle depth to two or three feet of water. You'll probably find a shallow trough just off the edge of the beach and in that trough many good fossils are lying waiting for a storm to dig them up and lift them to shore. The waves, that day, must be practically nonexistent in order to see through the water which should be clear. A bright sky will help and, if your feet are in soled shoes, your eyes and mind will work a lot better. If you don't have to worry about slipping or stepping on a sharp piece of coral or a skate or sting ray (which won't attack but which can be lying there partially buried in the sand) or a crab (which needs no explanation) you can look and see without interruption. Whenever possible, walk with the sun behind you and shining on the water. You all know about diving in deeper water for the really big and perfect bones, but this book is for the land-locked or bathing suit amateur collector, though the pictures will help the former, also. When there are no big storms and the bones cannot make it up onto the beach, it is amazing how close to the edge of dry land they can lie.

Usually it is more fun to hunt with somebody else. It takes some of the joy from a good find if you can't show it to a friend. There are manners in all things and this includes hunting for fossils. Whether you go with people who know no more than you do, or with experienced amateurs, or with a professional scientist, give them all room! Do not hunt right beside them or right in back of them. Certainly, stay with them the first couple of hours until you are fairly sure of what a bone looks like — then take off. There is plenty of beach. Stay within calling distance or plan to meet each hour. If you are talking, you can't really concentrate on hunting. If you are with someone who can help to identify your finds, do your collecting first, and then ask for help when you are both finished for the day. There is nothing more irritating to anyone who is bone hunting than to be shown everything that you find at thirty second intervals. However, if you find something pretty special, all beach manners are tossed to whatever wind that blows and the shout that accompanies the find will be heeded happily and hastily by any companion. It's so important to so many that it is found, that no worthy amateur ever resents your good fortune.

If you will come out of the water and look along the edge of the piles of shells thrown up by the high tides, you will find it most profitable. It is here that you will be most likely to find horse and deer, bison and capybara and tapir teeth (if you're lucky.) There may be various fishes' teeth, such as those from the barracuda, as well as fossilized crab claws and lots of sting ray barbs and the plates from their mouths. It is hard to find the complete pavement plate from the mouth of the ray but the "bars" that compose it are numerous.

Another, and even more rare and fascinating mouth plate, is that of the bony fish (vs. the cartilaginous varieties) of long ago. The parrot fish and puffer fish still use them. This can be found in its entirety, though usually it is not. These fish are so equipped because they eat the tiny sea life off the submarine rocks and corals. Teeth, as you know them, would be easily broken and thoroughly impractical. With these plates the fish can cling to their feeding spots and eat at the same time without the currents in the sea disturbing them. Nature is basic and realistic in its provisions. The plates are small and their patterns are delicately lined and interesting.

To continue with more plates, though not in mouths, study the pictures of the protective outer armor of the giant armadillo and his relative, the even bigger glyptodont. The glyptodont plates resemble six sided rosettes. The glyptodonts are now extinct. Armadillos still live but the modern one is tiny in comparison to his extinct relatives. The average size of one of the modern skin plates is about the size of the end of a middle finger. The plates that you can find of one of the old ones are seldom smaller in width and length than one by one and a half inches. Armadillo armor comes in shapes that differ in size and in the number of sides. These plates are called "scutes". The glyptodont scute has been described by one amateur as looking like a fossilized daisy. It does.

Alligators and crocodiles have lumpy plates under their skins, unlike the outer armor mentioned above. These, too, can be found. They peak toward the middle of a fairly circular piece of bone that is only slightly pyramidal in shape. It is covered with indentations which are in no certain pattern. The holes do not go all the way through the plate. The appearance is similar to that which results from heavy rain on sand — irregular sizes and shapes of shallow holes.

You will find multitudes of bits and pieces of fossil turtle shell at the water's edge, in the pebbles, and higher on the beach. They are easy to find, and the patterns on some of the shells resemble the figures and designs on Indian pottery. There is a great variety in the markings, and you will find pieces of shells of many kinds. The most common of these are the pond turtle, box turtle, gopher turtle, soft shell turtle, sea turtle and the giant tortoise. Each can be identified by the pattern on his shell. Whenever you find broken pieces of bone that have clear and well marked patterns, and they are not scutes from glyptodonts or armadillos, they will probably be parts of turtles' shells.

CHAPTER VII

There are also fossilized invertebrates — those organisms that have no spinal columns or backbones such as shells, corals, sand dollars, etc.

We will devote only a short space to the subject of ancient shells, etc. They are a different category of life entirely — different from all the things that we have been discussing. They are included here because they can be found on some beaches where there are no bones at all. These fossils are well worn by the sea and by their great age before they reached the beach. They may be many millions of years older than any of the animal bones. Their colors for the most part are gone, though their shapes and details are often amazing well preserved. These ancient fossils are much heavier than modern shells. You will find conchs, clams, oysters, pectens, tulips, arcs, snails, drills, cones, worm cases, many varieties that have no modern descendants, many that have no common names, and corals. Some of these things could date back three hundred million years or more though it is doubtful that these do. The oldest ones have been buried in some inland deposit since the sea receded or the land rose from it.

You can have a most interesting time looking for the "molds" of ancient shells. The external mold is the impression of a shell (or of a fossil bone) that lay on something soft and in which it was held and which hardened over the years. The shell, or bone, broke or wore away and a mold remained. You can think of it as a shell print and some appear to be as clear and as perfect as finger prints.

Then there is the cast. This is made when sediment handy to the shell seeps into it and fills each crack, crevice and turn. When time and the sea destroy the actual shell, the cast, which is by then solid and as hard as a rock remains.

This guide is concerned with the vertebrate world where spinal columns and backbones were the order of the day. The study and understanding of the invertebrate world is of tremendous importance and interest, as many of these animals developed and lived long before there were any vertebrates at all.

This division of organic life merits whole books of its own — and has them.

CHAPTER VIII

These, then, are some of the specimens that can be found in an amazing state of preservation. In those days things were really built to last. Fossil bones and teeth are shaped like modern bones and teeth. You don't need to dream up a shape for them. There are many bones in each animal and few are found in their entirety. Know what is needed in order to have a bone identified and you have won a great victory.

Wherever you find fossils on the beach you will become confused, at first, with other deposits and formations that resemble bones to the untutored eye. They are not. They are rocks or clay or asphalt or coral or anything else that the sea has tossed up on the shore that isn't bone.

Fossilized bone looks like rock, but it has a distinction that all collectors must learn.

Look for the tiny holes at the breaks or in the crevices. These were once filled with the nerves, marrow, and blood vessels of the live animal. No other formation has holes just like these. They look like holes in an extremely fine sponge — thoroughly irregular and covering the greater part of any broken surface. Some ribs that we have already described are dense and solid — no holes. Others are not. There will be no tiny porosites in teeth or ivory or enamel or ray and bony fish mouth plates or crab claws or ear bones. Most other bones will have them.

When you begin to hunt fossils on beaches you will be discouraged by finding so many bone fragments that you must realize cannot be identified as anything other than "just bone". Throw these away! They are too heavy in conjunction with your good finds to carry. Try to be selective in what you keep. You will see innumerable bits and pieces of ribs, few of which can be identified as other than, "it's a rib!" Of what? Well, you can't be certain, but if it's small, it can be rodent or fish material. Next in size are the dogs, deer, cats, etc., that were there. The larger ones could be from camels or horses or porpoises. Larger still might be from bison, and, if they get really big, they could be from the sea-cows, whales and elephants. The majority of ribs that you will find on beaches are from the animals that lived in the sea during the Pleistocene epoch.

On your first walk you may find the core of a horse's hoof that looks much like the hoofs used by today's horse. You can tell a deer's antler without help. Bending over will tire you at first, but continue to flip over those black or brown or gray rounded humps that are half buried in the sand. Nine out of ten will be broken rib fragments, but the tenth will almost surely be part of the inner ear bone of a whale. Porpoise ear bones are only an inch or so long and more delicately moulded than those of his sturdy relative, the whale. The part of the inner ear bone of a whale that you will find most often resembles the shape of a human ear that has been slightly "cauliflowered".

Here's to a bright day after a windy night — a calm sea and a strong pail — a cheerful though distant companion and lots of fossils on the beach.

It will be low tide again soon so hurry, and GOOD HUNTING!

ELEPHANT MATERIAL

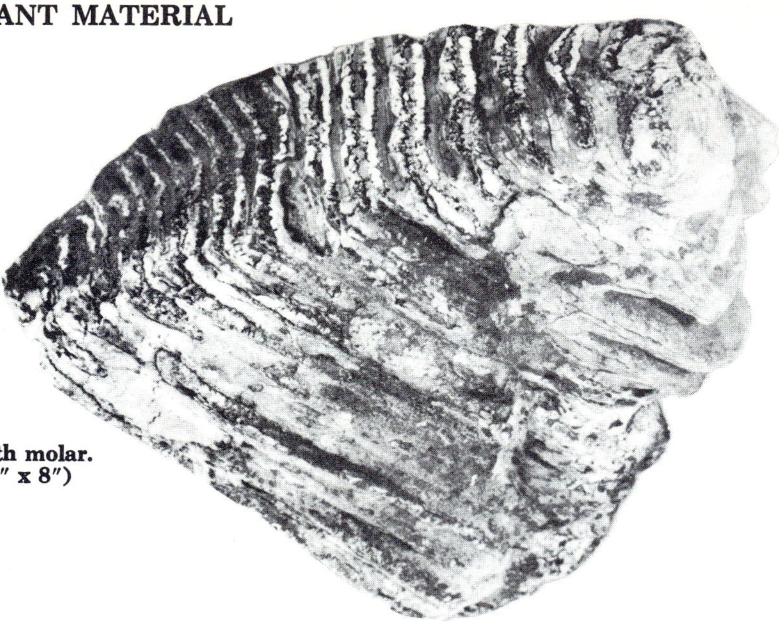

Mammoth molar.
(8″ x 3″ x 8″)

Mammoth molar.
(3″ x 4″ x 3″)

Mammoth milk tooth.
(½″ x 1″)

Mammoth milk tooth.
(½″ x 1¾″)

All dimensions given in the illustrations are approximate and are of those of the particular specimens shown. This is not necessarily true of what you find.

(2″ x 3″)

Piece of ivory tusk, one end polished. Note how it breaks in layers.
(8″ x 3½″)

Pieces of ivory tusk.
Note growth rings.

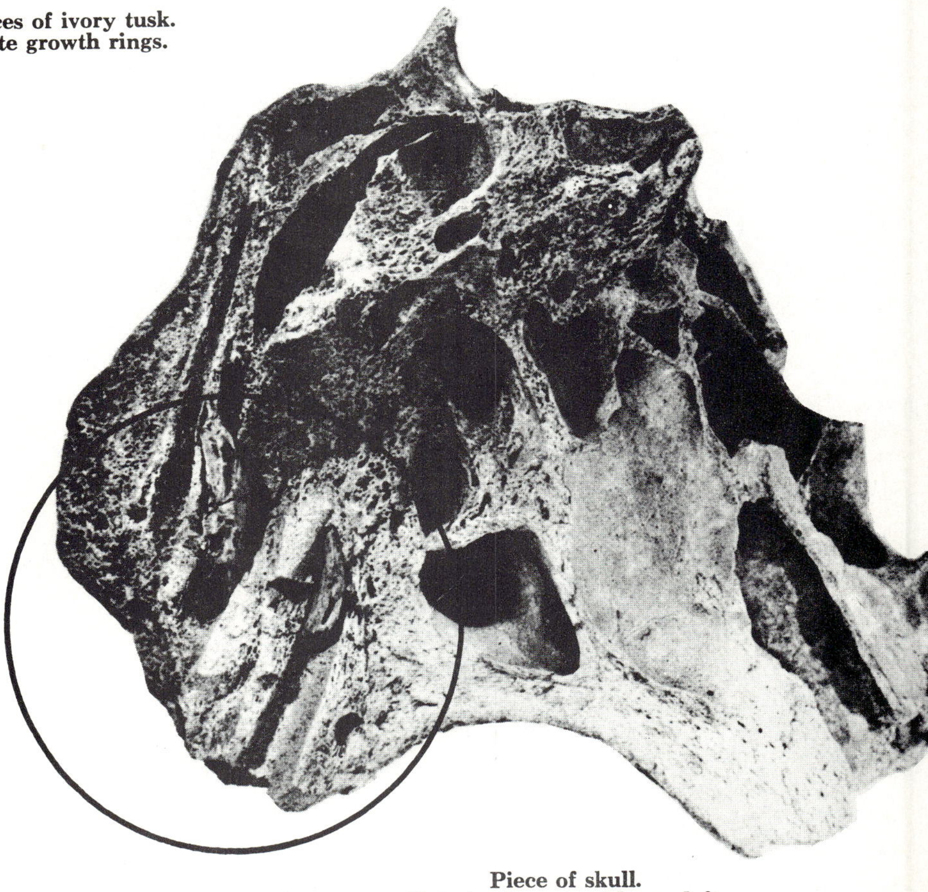

Piece of skull.
Note tooth roots at lower left.
(6″ x 5″)

Part of a vertebra.
(3″ x 3½″ x 2″)

Part of leg bone.
(8″ wide at top x 18″ long)

Mastodon tooth (5″ x 6″ x 3″)

(2½″ x 1½″)

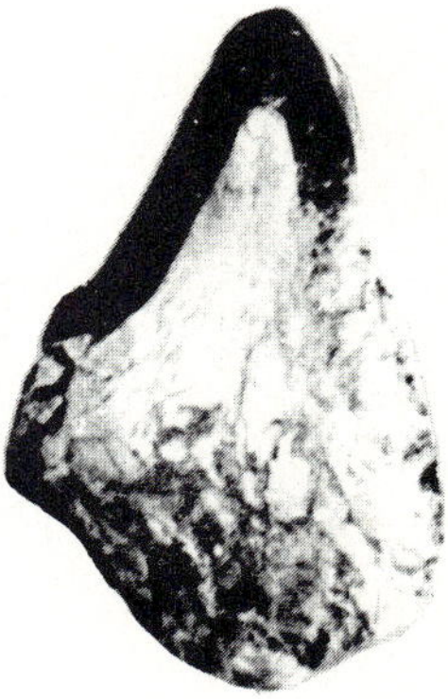

(2¼″ x 1½″)

Enamel cusps broken from Mastodon tooth. Note typical rim of enamel which aids in identification.

BISON MATERIAL

Two of the premolars. (1″ x 1½″)

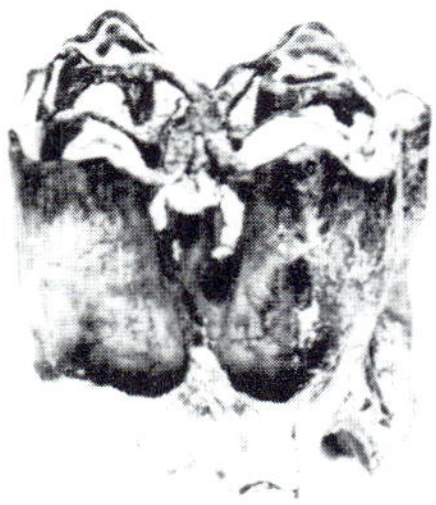

Molar, showing masticating surface. (2¼″ x 1¼″ x 1″)

Molar (2¼″ x 1¼″)

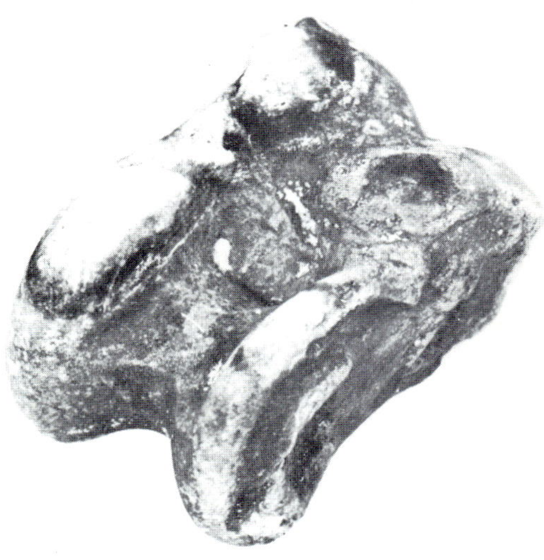

Astragalus (ankle bone) (2″ x 3″ x 2″)

End of back leg bone.
(8″ x 3″)

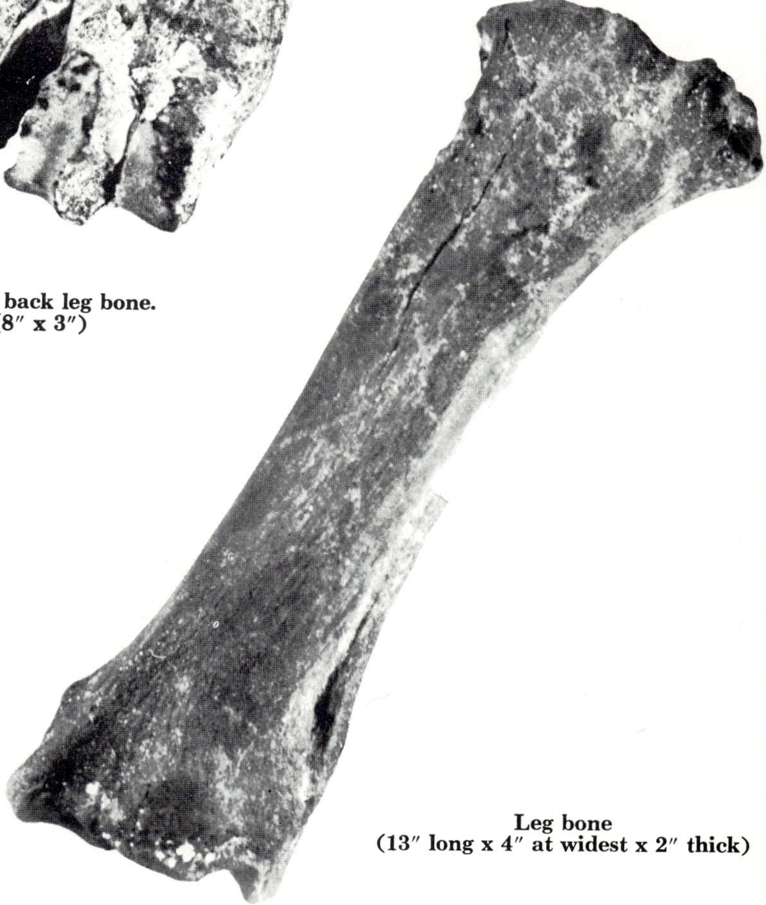

Leg bone
(13″ long x 4″ at widest x 2″ thick)

HORSE MATERIAL

Equus
Incisor
Masticating surface at the top.
(1½″ x ½″)

Equus
Newly erupted upper cheek tooth.
(1¾″ x 1¼″)

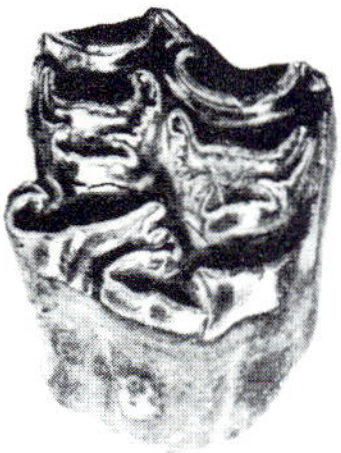

Equus
Molar
(1¼″ x 1″)

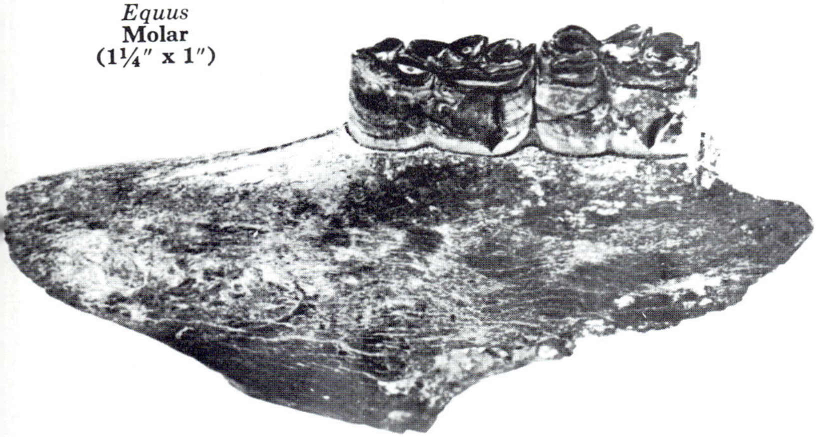

Equus **teeth in a section of jaw.**
(5″ x 2½″)

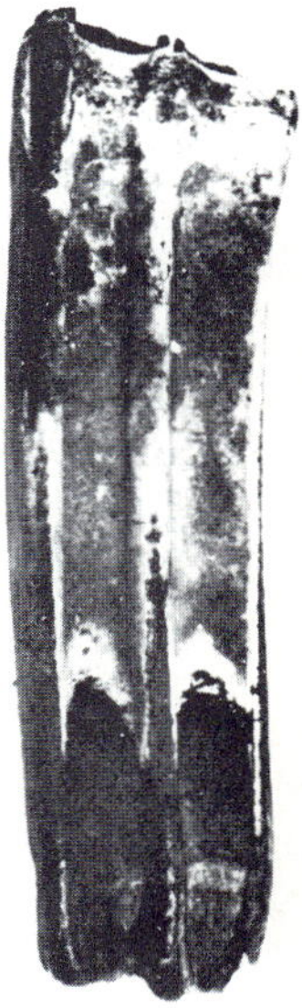

Side View,
Equus
molar.
(1½″ x 3″)

Teeth from
Nannippus
Three-toed, older, smaller horse.
(½″ x 1¼″)

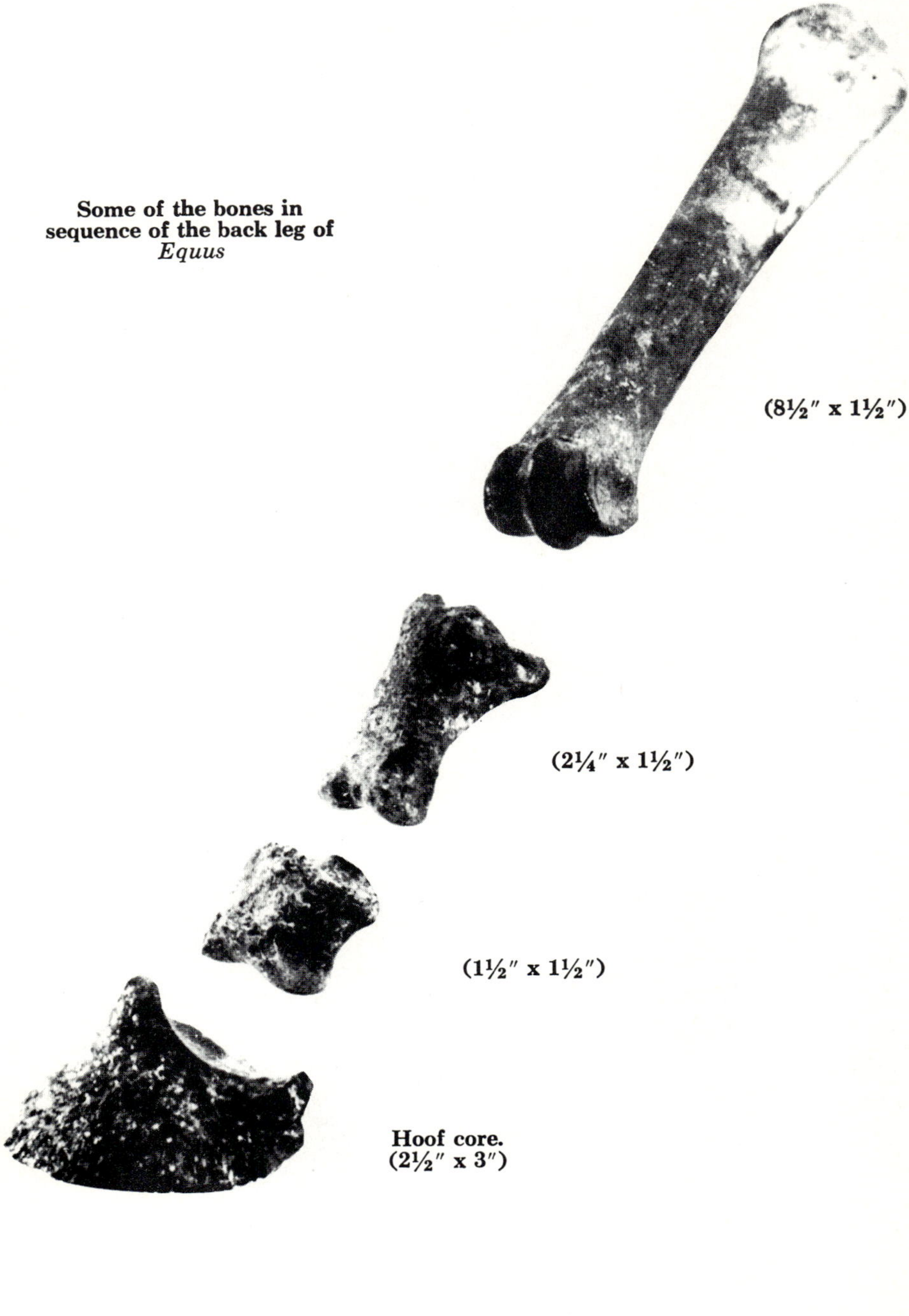

Some of the bones in sequence of the back leg of *Equus*

(8½″ x 1½″)

(2¼″ x 1½″)

(1½″ x 1½″)

**Hoof core.
(2½″ x 3″)**

Astragalus
(ankle bone)
Equus
(2½″ x 2½″)

Two views of an
Equus
hoof core
(2½″ x 2½″ x 2½″)

CAMEL MATERIAL

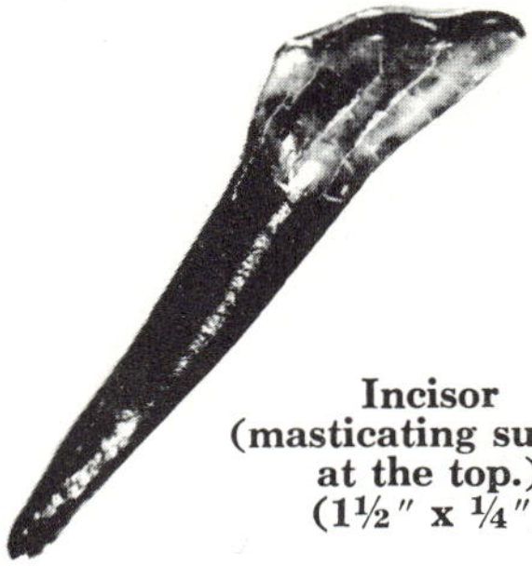

Incisor
(masticating surface
at the top.)
(1½″ x ¼″)

Premolar
(¾″ x 1″)

Canine (1½″ x ½″)

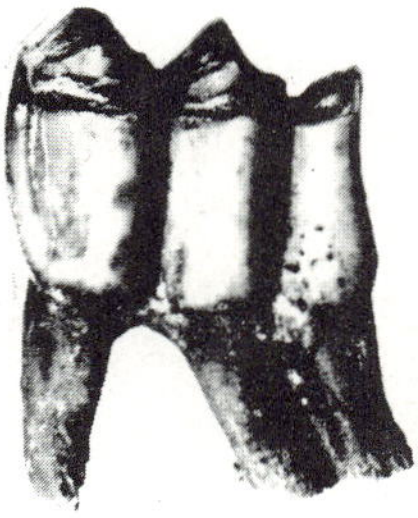

Molar (1¼″ x 1¾″)

Premolar
(1″ x 1″)

Toe bone
(2¾″ x 1″)

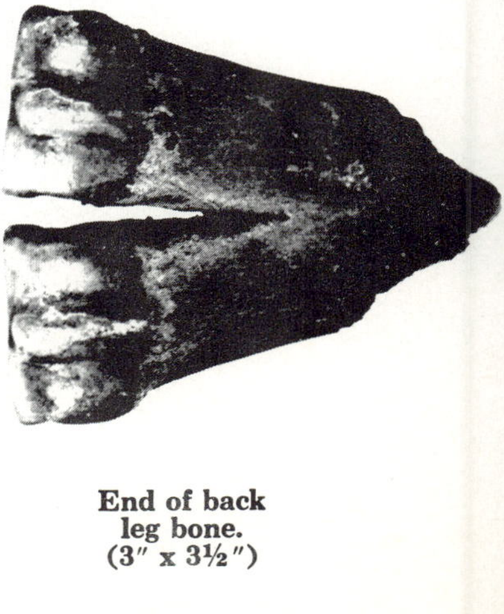

End of back
leg bone.
(3″ x 3½″)

DEER MATERIAL

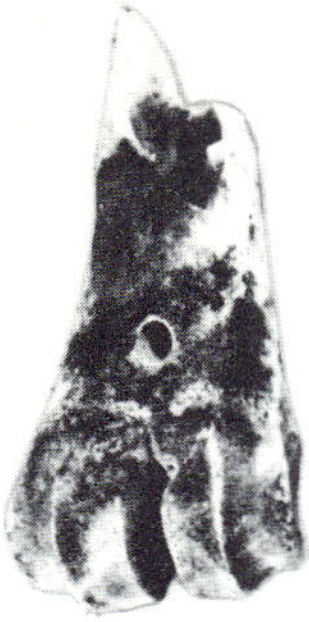

Part of back leg bone (1¼″ x 3″)

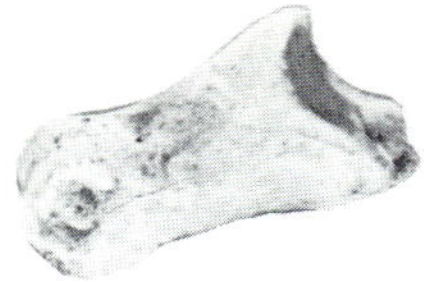

Foot bone (1⅛″ x ½″)

Teeth in jaw fragment (2″ x 1″)

Astragalus (ankle bone) (1¼″ x 1″)

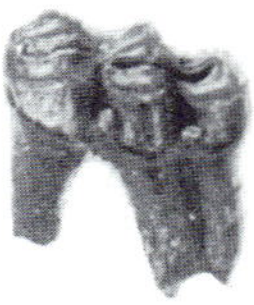

Molar (⅝″ x ⅝″)

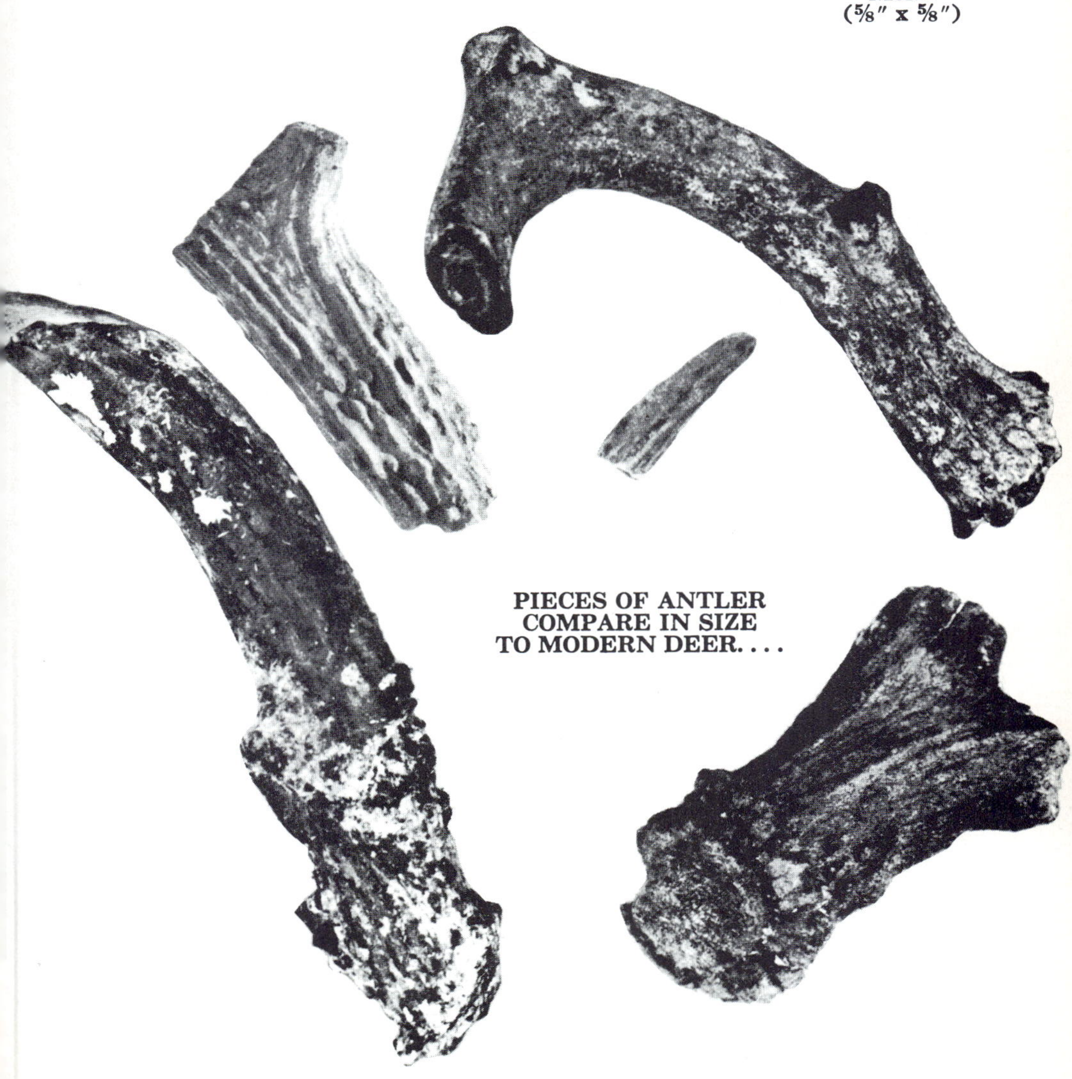

PIECES OF ANTLER COMPARE IN SIZE TO MODERN DEER. . . .

COMPARE FOSSIL WOLF TEETH WITH THOSE OF A LARGE MODERN DOG. NOTE THE SIMILARITIES. THE DOG JAWS HAVE THE WHITE TEETH.

Wolf tooth
(½″ x ¾″)

Wolf teeth
in fragment of jawbone
(2″ x 1″)

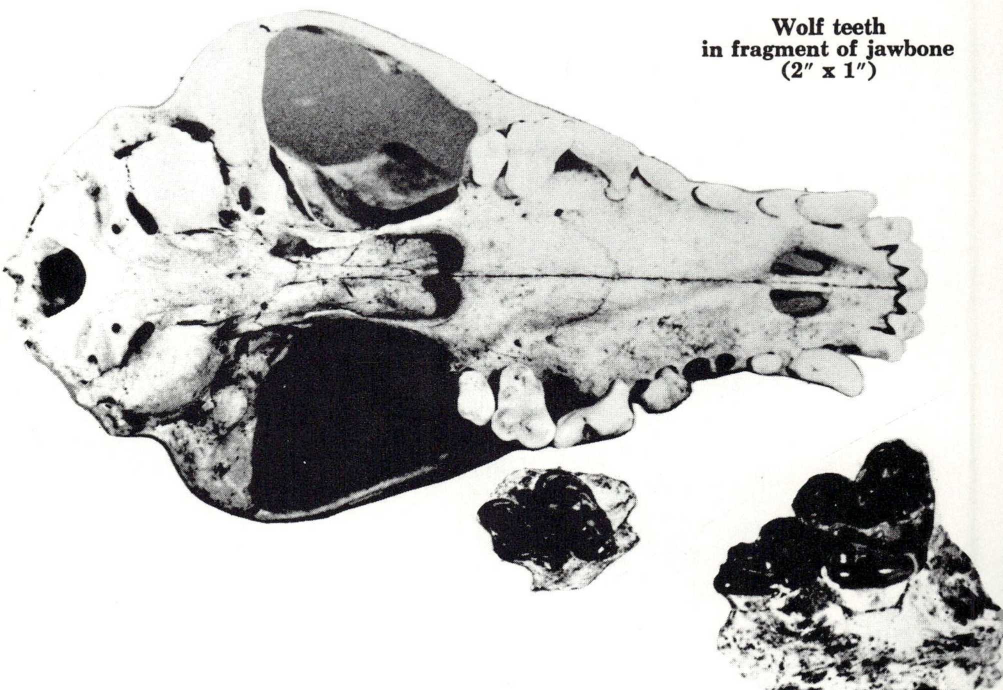

Wolf teeth
in fragments of jawbones
(Teeth: 1″ x 1″)

PECCARY MATERIAL

(3/8″ x 3/8″)

(5/8″ x 1/2″)

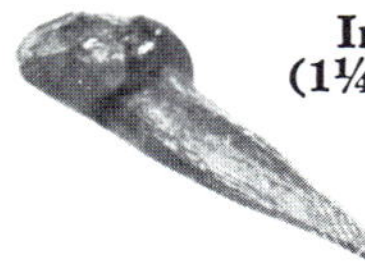

Incisor.
(1¼″ x ¼″)

(1″ x 1/2″)

Teeth of several species of Peccary.

SABER-TOOTHED CAT MATERIAL

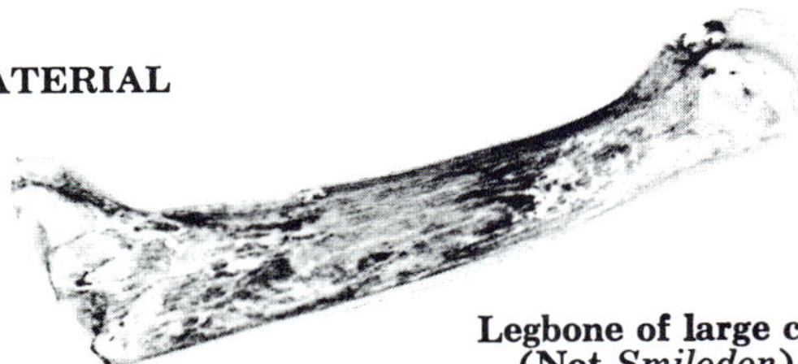

Legbone of large cat
(Not *Smilodon*)
(3¾″ x 7/8″)

Large cat canine
(Not *Smilodon*)
(2″ x 1/2″)

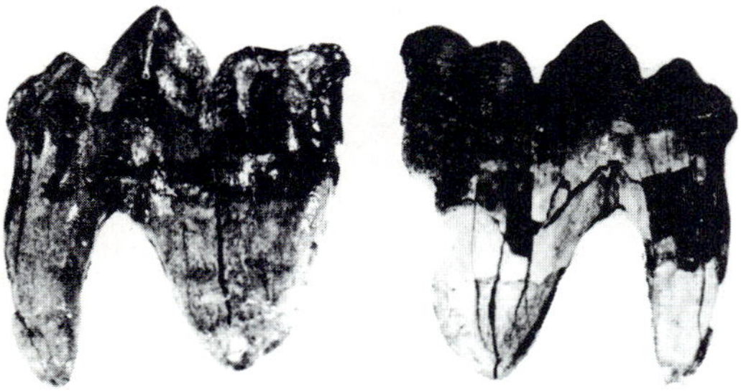

Back and front views of *Smilodon* tooth. This is from the largest cat, commonly known as "Saber-toothed Tiger."
(1⅝″ x 1¾″)

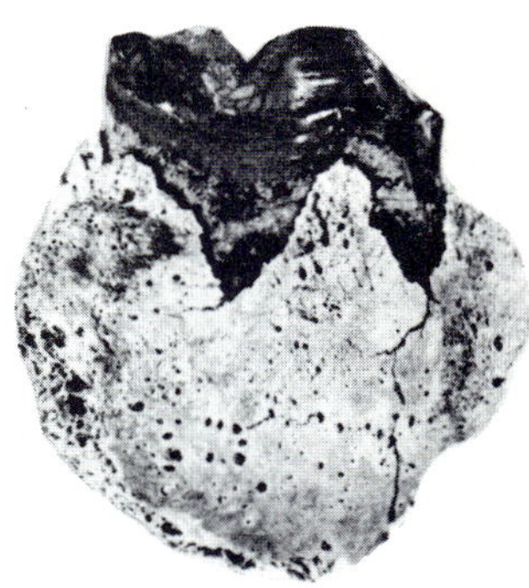

***Smilodon* tooth in fragment of jaw**
(2″ x 1¾″)

GLYPTODONT AND ARMADILLO MATERIAL

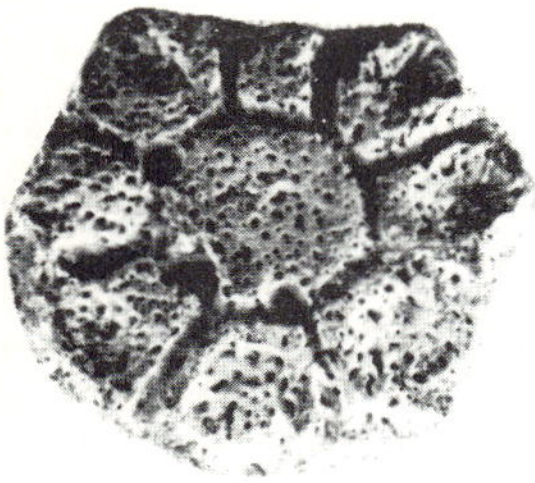

Glyptodont scute
(2½″ x 2″ x ⅝″)

Part of Glyptodont tooth. This specimen, rare and broken, has been made identifiable for you.
(1½″ x 1″)

Armadillo scutes
(2″ x 1″ x ¼″)

SLOTH MATERIAL

Two views of fragment and masticating surface of tooth of *Paramylodon.*
(1⅛″ x 1¾″ x ¼″)

Fragment of tooth of *Eremotherium*
(2″ x 1¼″ x ¾″)

Sloth claw (1½″ x 4½″)
(Not of Eremotherium)

RACCOON MATERIAL

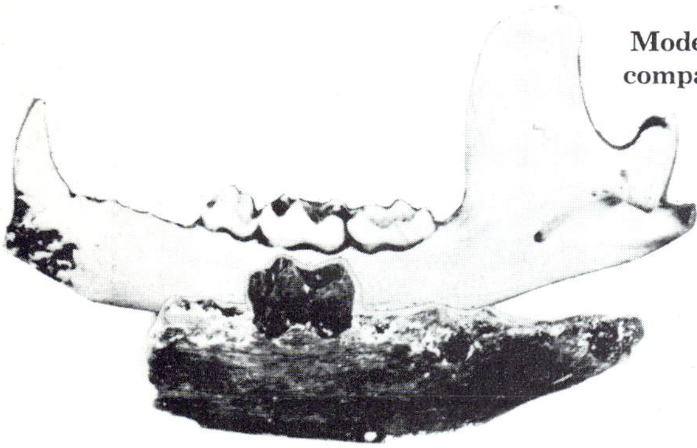

Modern Raccoon jaw with comparable fossil fragment.

(2″ x ¾″)

RHINOCEROS MATERIAL

Tooth

Top view

(1¼″ x 1″ x ¼″) ¼″ thick on biting surface

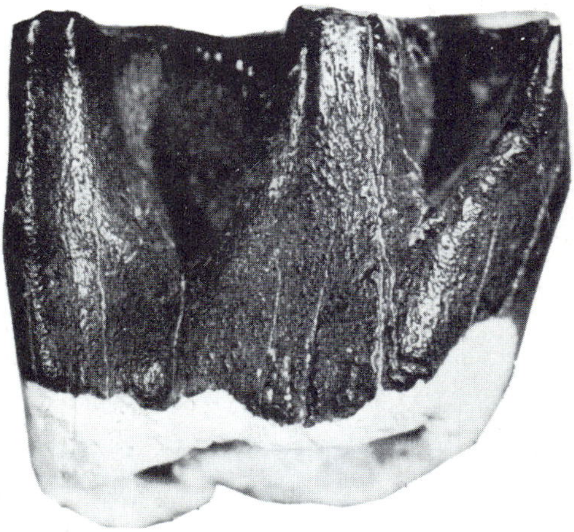

Side view

TURTLE AND TORTOISE MATERIAL

pond Turtle

pond Turtle

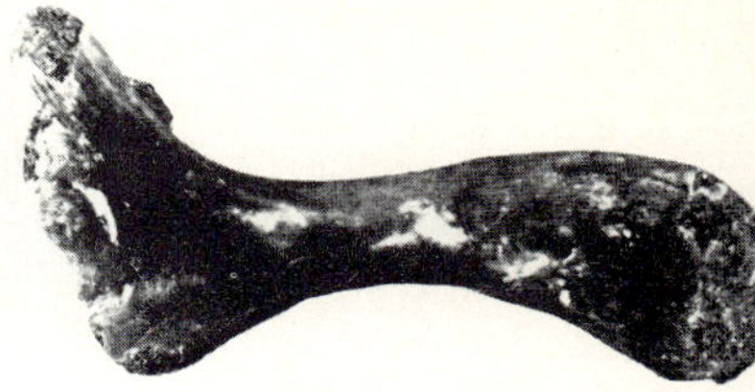

Leg bone (2½″ x 1″ at widest)

Fragments of several shells and patterns of various Turtles.

pond Turtle

soft shell Turtle

pond Turtle

Fragment of box Turtle still held in sedimentary rock in which it was buried.

Edge of box Turtle shell

Edge of giant Tortoise shell.

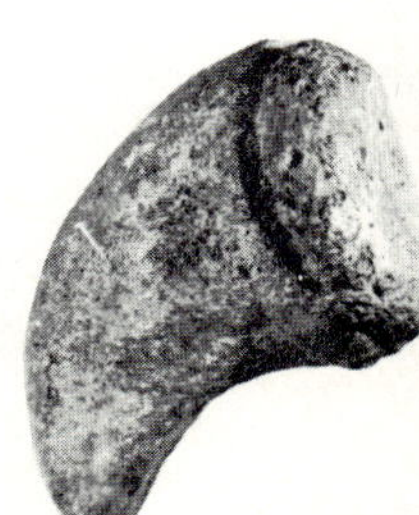

Spur from leg of giant Tortoise. (2″ x 1½″)

ALLIGATOR MATERIAL

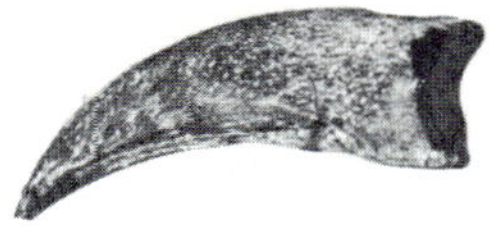

Claw core (1″ x ½″)

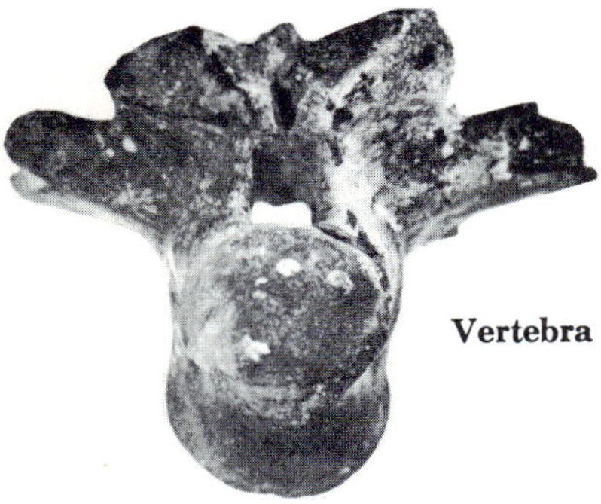

Vertebra (3″ x 2″)

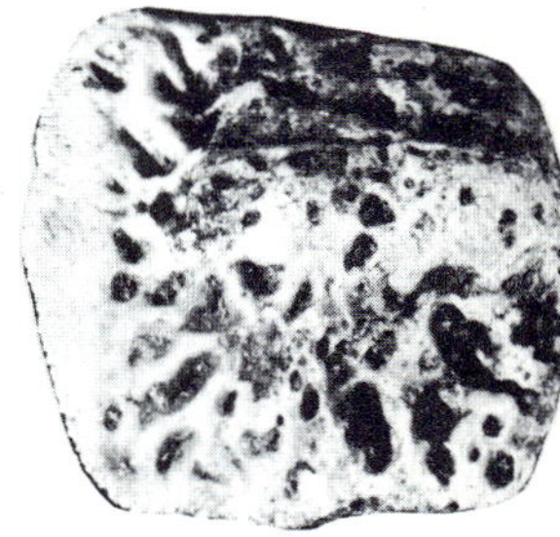

Under-skin scute
(2½″ x 1½″)

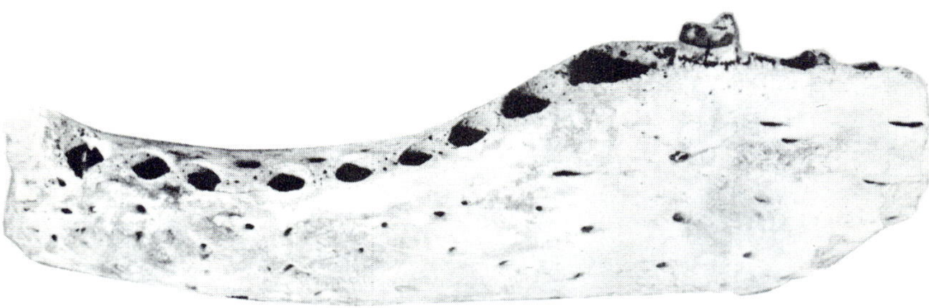

Fragment of jaw with teeth.
(5″ x 2″)

Teeth (average: 1″ x ½″)

WHALE MATERIAL

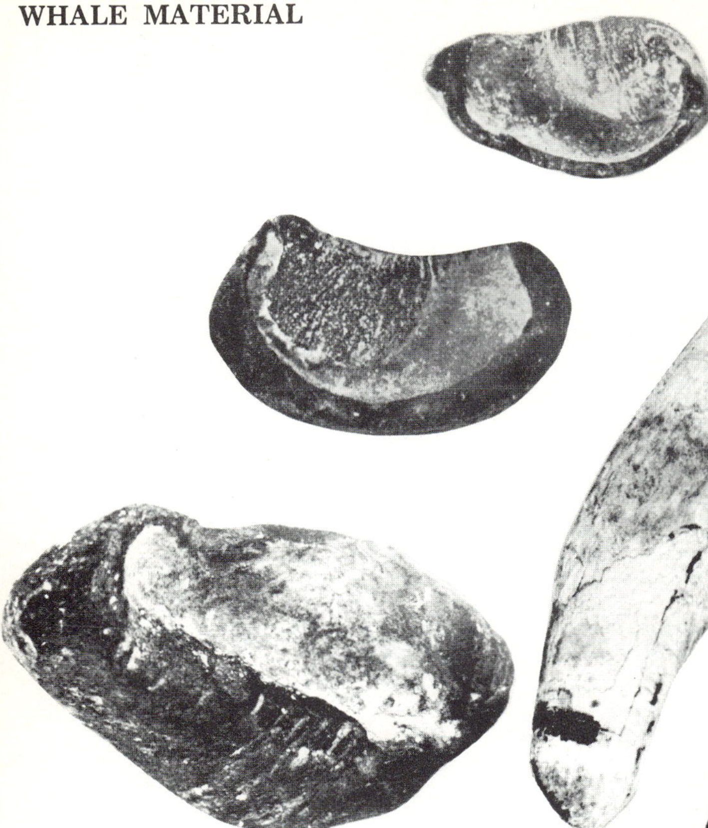

Parts of inner earbones of Whales. (Average: 3½″ x 2″)

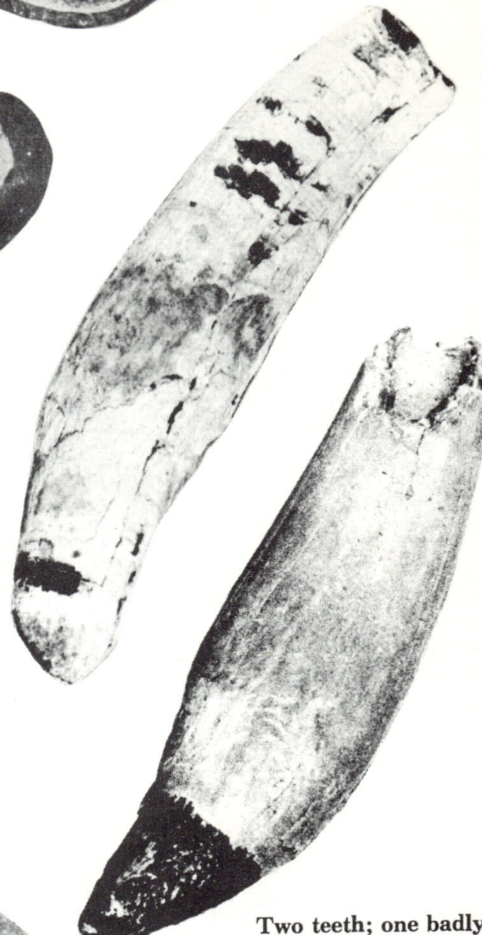

Two teeth; one badly worn, one in fine shape. (5″ x 1¼″)

Vertebra (8″ x 10″)

Part of vertebra
(2″ x 3″)

Small tooth in
fragment of jaw.
(Tooth: ¼″ x ½″)

Tooth (2½″ x ⅜″)

Note encircling lines for
identification purposes.

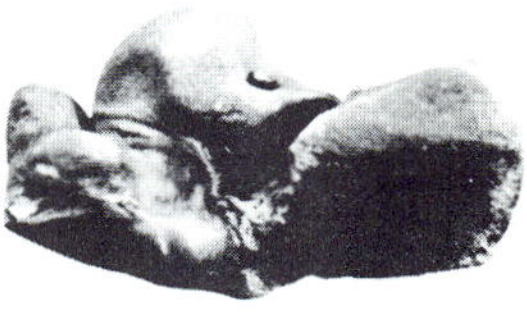

Inner earbone of
Porpoise.
(1½″ x ¾″)

It is unusual to find this bone so
complete.

Capybara molar (1″ x 2″)

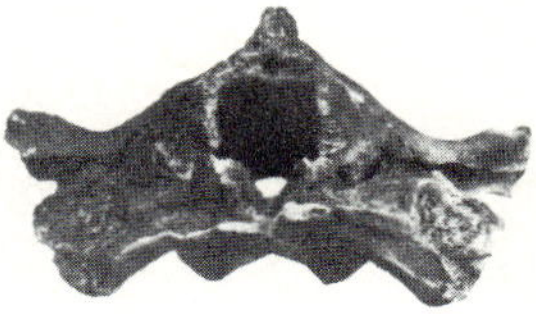

Part of Bull-frog skull (1¼″ x ½″)

Snake vertebra (½″ x ¾″)

Capybara incisor (½″ x 2½″)

Opposum teeth (¼″ x ½″)

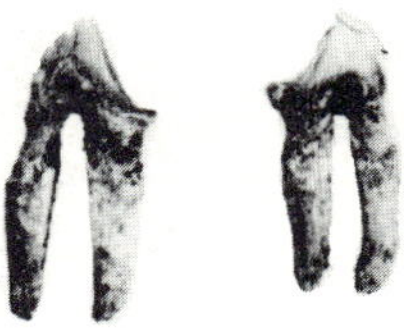

Tooth of striped Skunk in fragment of jaw. (⅜″ x ⅝″)

Tapir molar (1¼″ x 1″)

Rabbit teeth in jaw. (¾″ x 1¼″)

Rabbit incisor (⅛″ x ¾″)

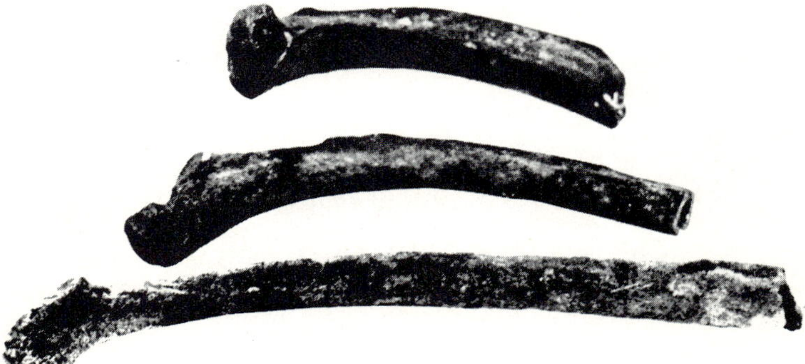

Animal rib fragments (largest: 8″ x ½″)

Two views of
giant Beaver tooth
(incisor)
(1″ x 2″)

Giant Beaver Tooth
(molar)
(½″ x 2¼″)

Bear molar
in the jaw. (¾″ x 1¼″)

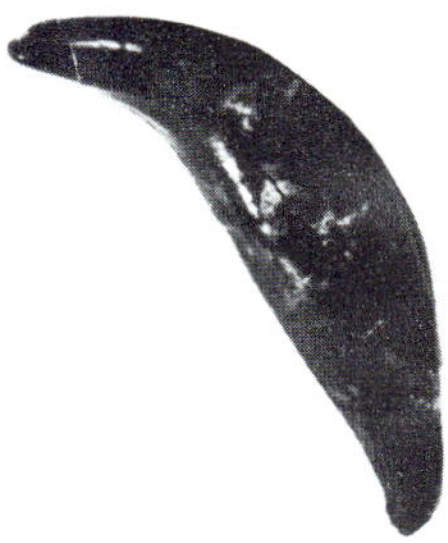

Bear canine
(½″ x 2¼″)

Bird bones
(¼″ x 2″)

SEA-COW MATERIAL

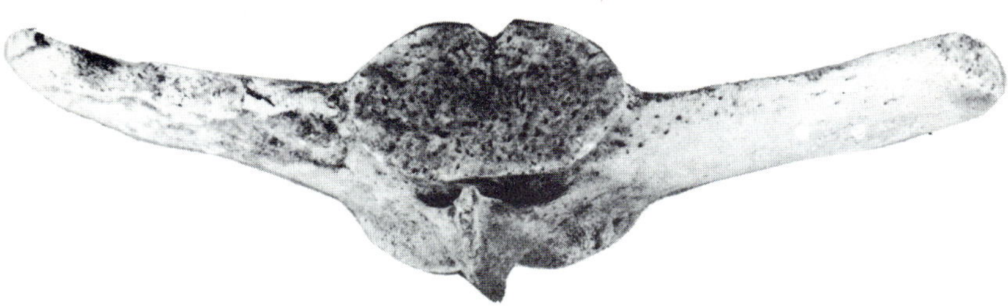

Vertebra
(8½″ x 2″)

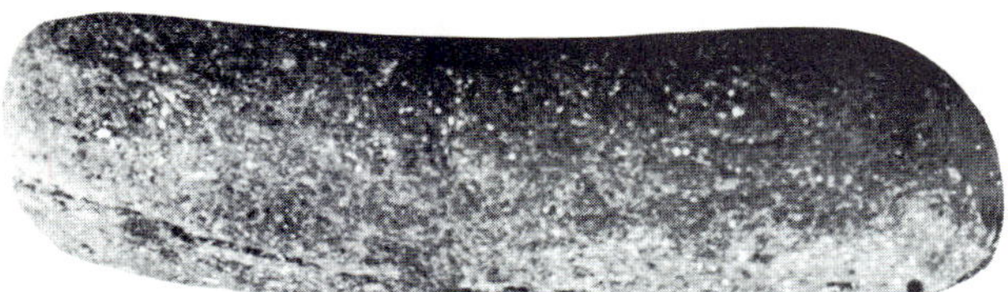

Piece of rib
(2″ x 5″)

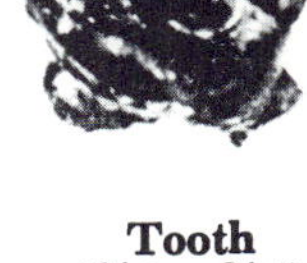

Tooth
(¾″ x ¾″)

Fish vertebra
(½″ x 1″)

Catfish spine
(1½″ x ¼″)

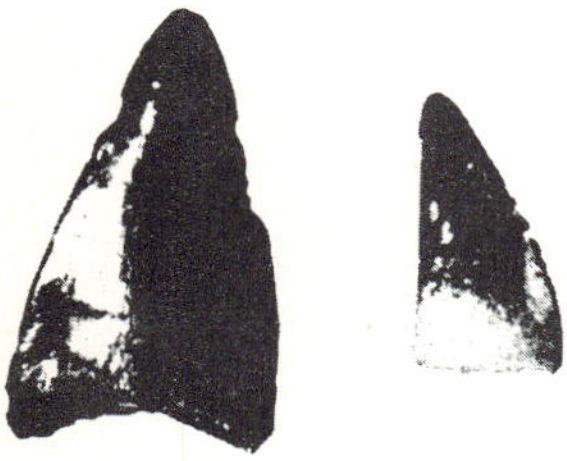

Barracuda teeth
(Average: ¼″ x ½″)

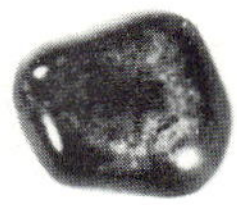

(¼″ x ¼″)

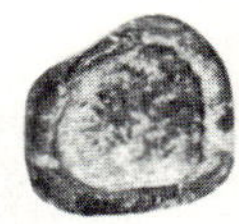

(1″ x ¾″)

Drum fish teeth
in and out of the jaw

Teeth from Sawfish bill
(⅛″ x ½″)

Mouth plates
of Parrot or
Puffer fish.

Gar fish scales
(Average: ½″ x 1″)

Crab claws
(Average: ½″ x 1″)

(Average: ¾″ x 1″)

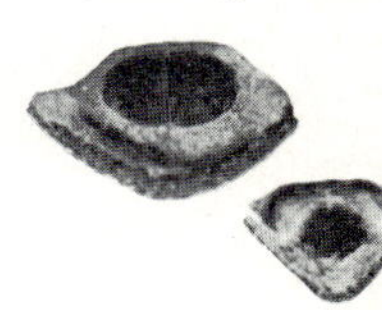

CARTILAGINOUS FISH MATERIAL

Shark spinal disk
(1″ x ½″)

Shark spinal disk embedded in coral (disk: 1½″ diameter)

Shark spinal disk
(1″ x ½″)

(1″ x ⅜″)

(1¾″ x ¾″)

Two dermal denticles (skin plates) of sting Rays. Photos greatly enlarged for identification purposes.

Sting Ray barb
(¼″ x 2¼″)

Bar of mouthplate of spotted Ray
(1″ x ¼″)

(¾″ x ¾″)

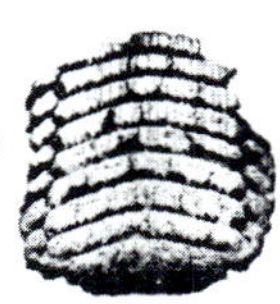

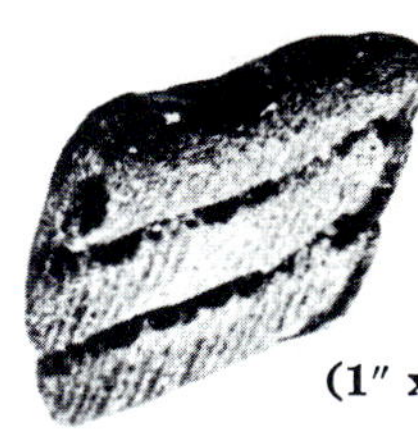

(1″ x 1¼″)

Pieces of mouth plates of sting Rays

Single bar of mouth plate of sting Ray
(¼″ x 1″)

Carcharodon megalodon—**no common name**

Family: Lamnidae—the mackerel sharks

This huge fish, previously mentioned in the text, is closely related to the living white shark, *Carcharodon carcharias.* The extant white shark has been recorded up to 36½ feet in length, however the majority that man has been strong enough to catch are only 8 to 10 feet long and are not full grown. These "man-eaters" live in all tropical and temperate waters. The only really good thing about them is that they are relatively few and far between.

The huge and heavy teeth of the fossil *C. megalodon* have serrations on both edges and are triangular in shape. You can usually recognize their shape even if the serrations have been worn off by the time you find them.

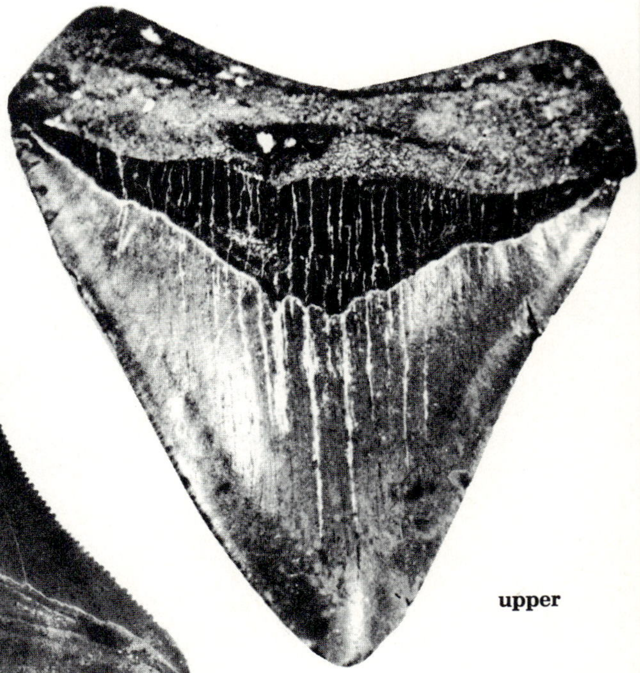

upper

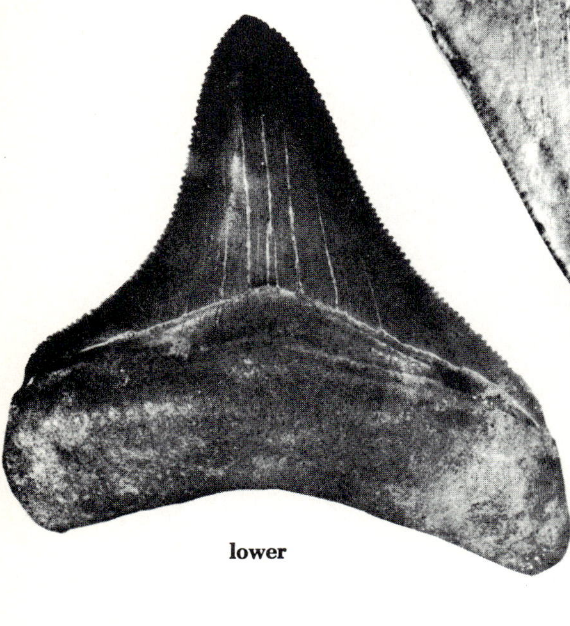

lower

The majority of the sharks' teeth pictured on these pages are larger than life size. However, there may be exceptions—both larger and smaller.

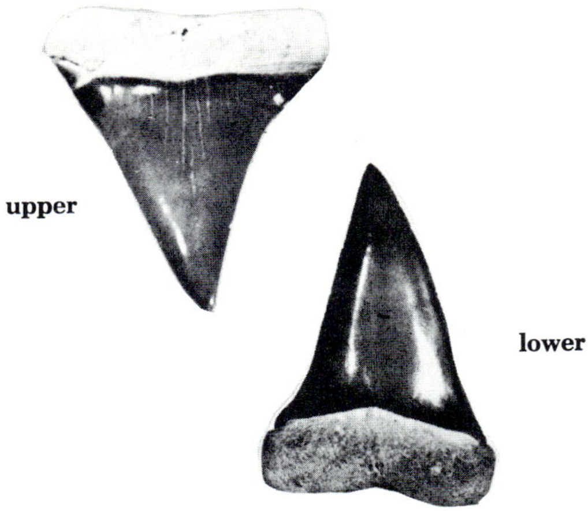

Isurus hastalis—**mako shark**

Family: Lamnidae—the mackerel sharks

This shark is extinct. Its fossil teeth resemble those of the extant mako shark, *Isurus oxyrhincus*, but they are much broader. The teeth of *I. hastalis* never bear the serrated edges that white shark (*Carchardon carcharias*) teeth always have. Even very worn white shark teeth may be distinguished from mako by the much greater bending of the mako.

Isurus oxyrhincus—**mako shark**

Family: Lamnidae—the mackerel sharks

Living mako shark, mentioned above.

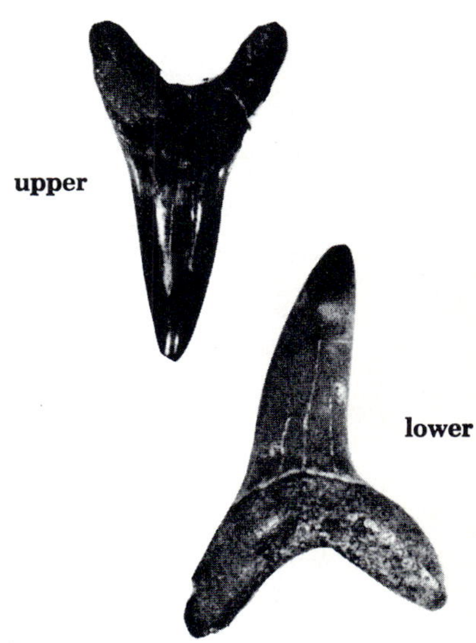

Odontaspis taurus—**sand shark**

(upper and lower teeth similar)

Family: Odontaspidae—the sand sharks

This species still lives. Its teeth are sometimes mistaken for those of the extant porbeagle shark that is a member of the Lamnidae (mackerel sharks) family. Teeth are found from ancestors of both these fish.

The sand shark, *O. taurus* has teeth that have broad trip-pod like bases. The porbeagle shark does not. The anterior and lateral teeth of the porbeagle are much shorter than comparative teeth of the sand shark.

Odontaspis cuspidata—**extinct sand shark**

(upper and lower teeth similar)

Family: Odontaspidae—the sand sharks

This extinct shark was apparently a large relative of the common sand shark, *O. taurus*.

Its teeth are heavier and bear more side cusps or denticles than do those of the living form. It also appears geologically earlier and was extinct before the beginning of the Pleistocene.

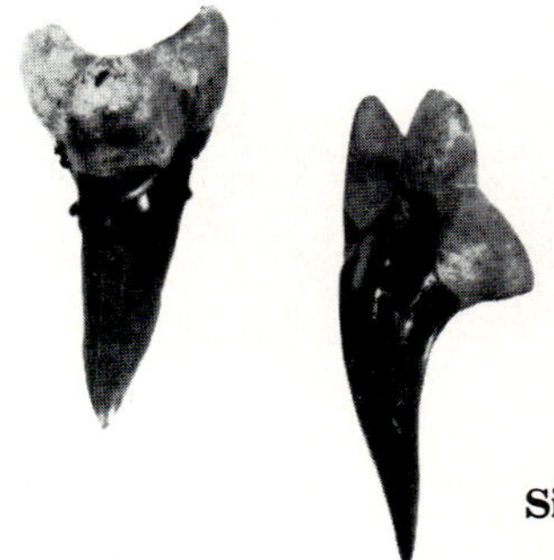

Two views

Galeocerdo cuvier—**tiger shark**

(upper and lower teeth similar)

Family: Carcharhinidae—the requiem sharks

This shark appears in the mid-Pliocene, the epoch before the Pleistocene, and the species is still living today. It is a scavenger and lists of the contents of the stomachs of the modern tiger sharks include such things as dogs, tar paper and sacks of coal.

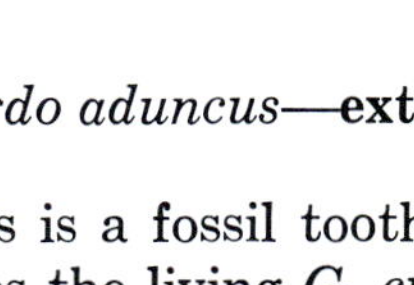

Galeocerdo aduncus—**extinct tiger shark**

(upper and lower teeth similar)

This is a fossil tooth from one of the extinct tiger sharks. It generally resembles the living *G. cuvier* but it is straighter crowned and generally less robust. *G. aduncus* disappeared by the mid-Pliocene.

Sphyrna mokarran—**hammerhead shark**

Family: Sphyrnidae—the hammerhead sharks

The hammerhead sharks are late developments, possible from carcharhinid stock. *S. mokarran* is the only hammerhead shark with serrate teeth.

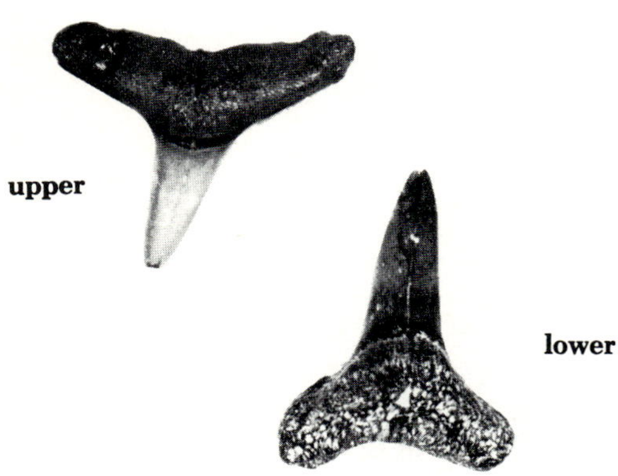

Negaprion brevirostris—**lemon shark**

Family: Carcharhinidae—the requiem sharks

Teeth of this species of sharks are distinguished by the combination of a long smooth-edged crown and a "wavey" edged base in the upper teeth. The lemon shark is still living.

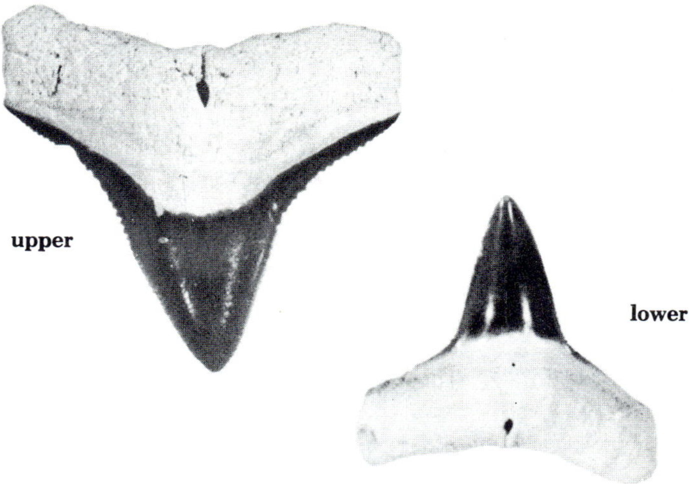

A typical *Carcharhinus*

Family: Carcharhinidae—the requiem sharks

There are numerous living members of the genus *Carcharhinus*. Basically they all have broad, serrate edged upper teeth and narrow, less serrate or smooth lower teeth. Many appear as fossils prior to the Pleistocene epoch and a minority have become extinct.

There are several different sharks in this particular group from which fossil teeth can be found. They include the spinner, bull, black tip, sand-bar (or brown) and the dusky sharks.

Only a professional who has made an intensive study of sharks can recognize the differences between the teeth of this genus and determine their identifications accurately.

Hemipristis serra—**no common name**

Family: Carcharhinidae—the requiem sharks

This shark is extinct in the new world, but still lives in Eastern Asian waters.

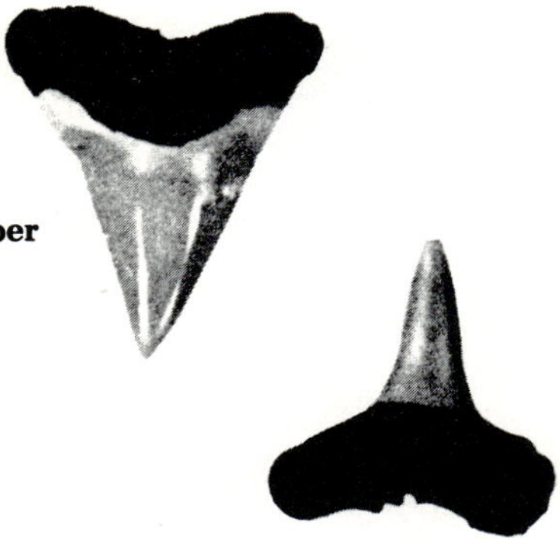

Carcharhinus leucas—**bull shark**

Family: Carcharhinidae—the requiem sharks

The bull shark is the only shark frequently living in fresh water. It is also common in shallow marine situations such as bays and river mouths.

Family: Hexanchidae—the cow sharks

These are rare fossils and difficult to find. Search the beaches at the foot of the Calvert and Scientist Cliffs in Maryland. As is noted in the text, one must have many teeth for comparison in order to positively determine the specific name of the fish. Because of the unavailability of this quantity, it is impossible to say from which particular kind of cow shark the illustrated teeth came.

Cow shark teeth are easy to distinguish as they are composed of stair-stepped cusps unlike those of other sharks.

There are both extinct and extant species. Most of the extant cow sharks are now living in deep water.

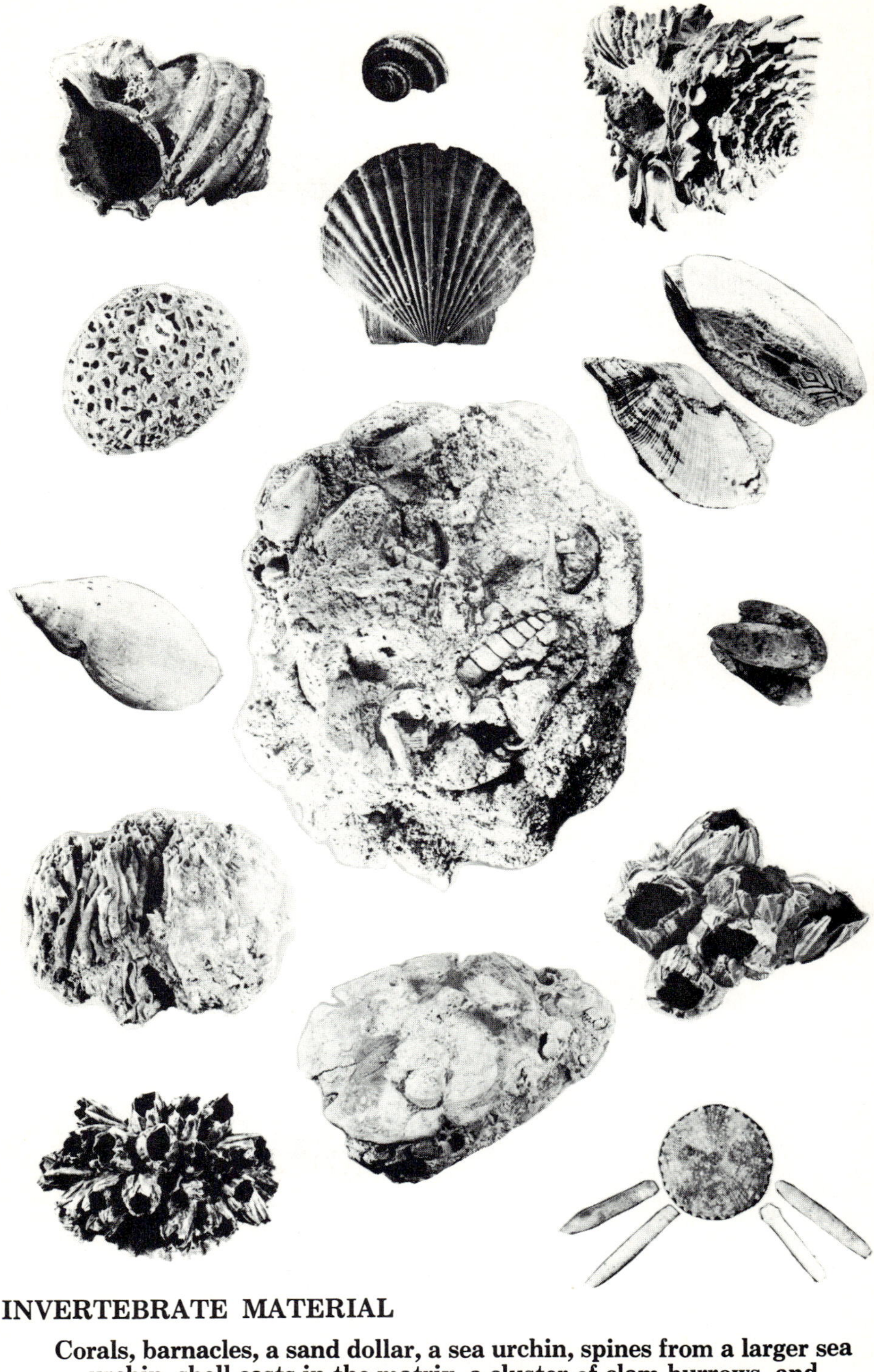

INVERTEBRATE MATERIAL

Corals, barnacles, a sand dollar, a sea urchin, spines from a larger sea urchin, shell casts in the matrix, a cluster of clam burrows, and ancient shells.

Typical spoil banks formed as the result of the excavation of a canal.

A "flat" with more spoil banks in the background.

PART II

. . . AND BANK

CHAPTER I

Banks are safe places for valuables that should not be left in the house. However, some "banks" contain valuables that should be removed for their own safety and preservation.

It is in the latter that you are interested. No specific sites can be listed for reasons that will soon become obvious.

As you walk or drive near many of the fossiliferous beaches that bound parts of the continental United States, you will see much earth moving work. Builders are clearing land for construction. Engineers are laying beds for highways. Farmers are digging drainage ditches. Mining companies are digging pits in which to find various kinds of minerals. Land development companies are providing canals and lakes for their properties which can then be called "water front".

All these activities performed by various sizes, shapes and powers of bulldozers, dredges and draglines reach a common juncture in the course of their work — the earth that is the residue of the necessary removal is dumped in piles. These piles are called "spoil banks". On many of these banks can be found those elusive but fascinating remains of the past — vertebrate fossils, that you learned about in Part I.

The sites will be near enough to sea level so that machine digging will reach the fossil bearing strata and expose, on the spoil areas, fossils similar to those that you have found on the beaches. It will depend on how sharply the land rises from the beach as to how far inland you will be able to find fossiliferous banks. If the rise is gentle, and if it levels off fairly soon, you can hunt a good way inland. If the rise is in the form of a cliff, hunt in the sides of the cliff or stay on the beach. You are interested here in the slow rise.

The aforementioned pits for minerals will be dug the deepest while drainage ditches may be only a few feet deep. The high spoil banks caused by the deep excavations for pits will ordinarily produce fossils in areas at greater elevations above sea level than will ditches. Banks that result from digging canals fall between the two. The foundations for buildings call for a depth commensurate with the proposed constructions' heights or weights. Road beds are the most shallow and will have the lowest piles of spoil in consequence. However, this does not necessarily mean that interesting and rare fossils cannot be found as a result of digging only a few feet below the surface. If there are vertebrate specimens in the area that you are hunting, one or another depth (or height) will bring them to light. It will be up to you to find the location that provides the necessary combinations.

While the blade of a bulldozer scoops the earth from the ground, the dredge differs in its working locations and methods of operation. Its usual job is to deepen or widen existing harbors, bays and rivers. The water cannot be separated from the earth that must be removed so some of it must go, too. The job is accomplished by "chewing" to the necessary depth or width and then pumping the waste material through huge pipelines and under great water pressure. The water, sand, mud, clay, shells and accompanying fossils are all spewed out onto dry land. All the machinery is mobile and slowly progresses to the desired length and width of the project. This operation leaves a line of continuous spoil banks. Each pile is necessarily as high as the depth to which the ground has been lowered. Individual mounds are made as the result of digging single holes. Some banks are four feet high — others may be forty.

Dragline operations are usually limited to digging canals, creeks, drainage ditches, and to performing open pit mining.

CHAPTER II

Fossil bones and teeth in the spoil may have lain undisturbed except for earth actions through the ages. Many thousands or hundreds of thousands of years have passed since their burial place was in either ocean, marsh, creek, gulf, river, or on land. Most have probably been dry and protected from disintegration by the lack of air. If the animal died there it may only now have been disassembled by the machinery. At least it has not been faced recently with waves and ocean currents and storms that broke it from its sedimentary bed, tumbled it over and over and crashed it onto a beach.

On the banks and protruding from them you can find sharks' teeth, some of which have been amazingly well preserved. The sea actions break them more often than do those of the earth or the machinery.

You will be able to see the differences betwen the average beach and bank fossilized specimens. The latter is usually far better preserved. The masticating surfaces of teeth are more clearly defined — less worn. Leg bones and vertebrae may be complete. Even delicate ribs may have the articulations at the end intact. While on the beach an ankle bone is a good find, spoil banks hold quite perfect foot and toe bones. On a sunny day with great patience and the eyes of an eagle, tiny fossils such as the small vertebrae of snakes, fish teeth still in the jaw, leg bones of bats, or a pocket gopher tooth can be reclaimed. You will be hunting for specimens that would almost certainly have been destroyed by the actions of the oceans.

When the pipes of the dredge are large enough and the masses of earth scooped by the bulldozer or dragline big enough, they will contain large, as well as small, fossiliferous remains. Here one can find such huge specimens as mammoth teeth still in a fragment of the jaw bone. An occasional mastodon joint or piece of skull or leg bone can be there. Bison teeth in a section of jaw and undamaged whale vertebrae are in some of the resulting piles. Saber-toothed cat bones and teeth and giant sloths' claws may come through undamaged. You can stand on one of the huge pipes and hear the spoil hitting on the sides as it rushes through. One never ceases to be amazed that it isn't ground to bits in the process — some is, but a lot makes the trip in perfect shape.

Lack of space in this book precludes photos of everything that can be found on either beach or bank. When one considers that each animal has many bones and teeth it is not difficult to understand why only a bare minimum are shown. As you hunt both beaches and spoil banks you will soon be able to determine which illustrated specimens came from which sites. The inclusion of worn material is done with purpose. There is no use believing that amateur collectors (or pros) find only museum display-type fossils. On both beaches and banks they are rarities. Those things pictured can be found with ease some places and with great difficulty, if at all, in others. Just remember that you may find a very worn bone or tooth that can still be of great scientific interest because it is so rare. No collector worth his salt will ever be so satisfied that he won't attempt to do ever better in the field. However, it is foolish to be dissatisfied with many finds that are not as complete and perfect as are new skeletal remains.

CHAPTER III

There is yet another area in which fossils can be found. In some locations there are extended expanses of earth that have been cleared of trees, grass and top soil. For trailer parks or highways, as examples, there is no need to dig deep into the earth. At such cites the ground is only removed to the depth necessary for a firm and level foundation for construction. These places we call FLATS. A flat is also made by a dredge that spills its deposits on sloping ground so that the water carries the residue over several acres rather than piling it. These are great places to hunt as the walking is much easier and there is little climbing.

When the topsoil of a site has been removed by machinery — pushed off rather than scooped out to any great depth it becomes a flat. Here, there may be exceptional fossil finds. Unlike the individual bones and teeth that you can see while hunting on beaches or banks, one has a chance of finding on a flat the skeletal remains of a fossilized animal "in place". This means to the scientist and to the amateur collector, the place where the animal died. You will seldom, if ever, find one with the leg bone connected to the thigh bone or the thigh bone connected to an undisturbed line of vertebrae. Not only is there no soft material left to join the bones one to the other, but the animal has been there at least thousands of years. Because it lies close to the earth's surface, nature's elements from rain to tree roots to ice and to the ever shifting ground will have upset the structure. Sometimes it is found as a hopelessly inarticulated jumble of fragmented bones and teeth. Hopefully the time will come when you can see mammoth teeth in a piece of jaw bone, protruding sections of vertebrae and ends of leg bones near ribs either exposed or in suggestive earth ridges. When you walk (after the flat has been cleared of all vegetation) by the elephant site a few days after it has rained on the exposure, you will observe new low mounds as well as ridges. Here you can surmise there is additional proof of a fairly complete skeleton. The earth has sunk around the bones and teeth while they have stayed at their original depth in the ground.

With your hands, a brush or at the most a trowel, very gently uncover these hillocks to the extent of proving that they are covering fossils rather than lumps of clay or dirt. If they are, stop.

Naturally, as you are interested in the preservation of the ages and in working in a scientifically acceptable manner you have received permission, from either the owner of the land or the foreman of that particular job, to hunt on that property. As you have not invaded posted or fenced property without permission, go to the nearest pertinent authority, explain your find and ask that the site not be disturbed before you can get professional help. You will be assured that the bones will not be moved or removed. Engineers, drivers, and land owners are all very helpful people and are very necessary adjuncts to your hobby and work.

Call the university or museum nearest to the site and ask to speak with a vertebrate paleontologist.

DO NOT TRY TO EARN YOUR Ph.D. IN THE FIELD. YOU CAN'T!

If you should attempt to remove the fossilized skeleton of any animal larger than a small rodent you will destroy it. Even a rabbit would be too big. To remove a mouse, you would have to excavate the skeleton with the large clump of earth in which it is lying. Think of it as an extremely rare and precious plant that you are taking from the ground with the intention of transplanting it. You would have to work far away from all its sides and deep beneath its roots. So, too, must any small animal in place have a deep and wide ditch dug all around it until the entire matrix can be removed leaving the fossil undisturbed in it. Larger beasts are removed in sections or by individual bone and tooth. The professional knows how — the amateur collector does not. This would be impossible without the professional training, experience and the implements of the trade.

The paleontologist to whom you speak will first wish to be certain that you know a fossil when you see it. Many busy men have rushed great distances to aid in the removal of what turned out to be a cow that was alive ten years ago. They have to take a chance on your knowledge so don't cry wolf. The earth scientists are far too busy for this kind of fun and games. When they are persuaded that you know what you are talking about they will tell you when they will meet you at the site or send assistants to prepare the specimen and then to remove it. It will be much sooner than you had thought was possible.

Your job, in the meantime, is to make absolutely certain that the site is neither disturbed nor its location made known to anyone other than the owner of that land or to the aforementioned authority.

To say that the scientific value of, and interest in, any fossilized remains of an animal (other than the most rare) is commensurate with "its whole entirety" is redundant. Nevertheless it is said and with purpose. There have been numerous "entire" skeletons that were "whole" when you called the professional but were far from it by the time he got there. You must protect it from people, young and old, who want "just a little piece for a souvenir". Unfortunately, they are multitude in their thoughtless though well meaning ignorance.

If it is at all possible, cover as much as you can with tarps or blankets. Wind and rain, at this point, can be very destructive.

You will be amply rewarded, for all the miles and days that you have walked hunting with little success, when the paleontologist arrives. Not only will you be congratulated and thanked but you will be permitted, under direction, to help in the excavation of your find. You will be shown how "plaster jackets" are made. These prevent further damage to the fossil in the removal process and assure its safe arrival at the laboratory in which it will be restored and preserved and studied. You will be given written recognition and credit for your contribution to the knowledge of the past. This will be a great day for any amateur collector.

You will find much personal satisfaction in the fact that you have not been hunting in established fossil beds where the pros work. There they have learned by past experience where they have a reasonable chance of locating the remains of a bird or beast or fish in place. You have found something where no pro expected it to be.

CHAPTER IV

Bones and teeth will be, as are your beach finds, colors that result from various combinations of minerals, chemicals and vegetation in which they have been buried. If they are black and you have found them on banks or flats that are brown or tan, it is unlikely that you will find the whole animal. It is probable that those fragments are not in place. There has been as much turmoil in the earth as there has been on the surface. There, fossil remains can be moved slowly and to great distances. Time, rain and the constant movement of natural waters, both salt and fresh, account for many of the bits and pieces that you will find in what you consider to be the "wrong" place.

However, exceptions to this are many. One well known scientist has the jaw of an ancient horse. Three teeth in it are different colors — one black, one bluish and one reddish. Far stranger things have happened than finding fossilized skeletons whose colors do not match that of the spoil in which they lie. For example: It is pretty traumatic for an amateur collector to see five horse teeth within the radius of two feet, dig to locate the jawbone that certainly must be directly beneath them, and find instead, three well preserved whale vertebrae — no sign, whatsover, of the horse jaw. So it goes.

When you no longer fight the "impossibilities" that constantly confront you in this hobby, you will realize that the mysteries and surprises make it that much more interesting. As with other kinds of collecting, the longer you hunt, the more you will find. Your "inner-eye" will suggest to you that "just there" you'd better slow down and really pinpoint your attention on one particular section of bank or flat. With experience you will learn only to scan areas that you are quite certain will be unproductive.

When you hunt spoil banks that are beside the fresh or brackish waters, come back via the edge. Fossils are heavy and many roll down the banks. Use your stick or rake and scratch gently in the wet earth as you walk very slowly. It will be worth your time and effort.

You will work many spoil banks that produce fossils for the distance of a city block or two. Suddenly there are no fossils either high, in the middle or at the water's edge. Keep walking. If you're lucky fossil deposits will reappear just as suddenly as they disappeared. This is easily explained as you know that the project is dug to a uniform depth or scraped off evenly. However, the fossil strata need not conform to man's digging. The strata of earth in which animals lived and died may not parallel the present level of the earth's surface. You will find fossils where the excavator has run into pockets of them and where his levels and depths and theirs coincide. There are hills and valleys under the surface in which the machine works. Layers of earth and deposits therein do not necesarily lie in a straight line.

Sometimes you find fossils on the sides of banks but none on top. The operator of the dredge or scoop has worked deeper than that particular fossil stratum. Fossils are deep in the pile. Keep track of what happens after a heavy rain or wind. The surplus non-fossiliferous material may be washed or blown off. Don't go fossil hunting right after it rains. You'll be working in a sea of mud in which you can see nothing. Wait for two or three days until the earth has had time to settle. It settles around and beneath the bones and teeth and leaves them exposed. You will be able to hunt many inland locations more than once. Each good rain or windstorm uncovers fossils if they are there. Some fossil bearing strata are very wide — others are not.

Always send bones or teeth of indeterminate age or species to the pros for identification accompanied by some of the sedimentary deposit that surrounded the specimen. Unlike beach fossils, those from banks or flats are not necessarily divorced from their pasts. The earth in which they have lain may enable the paleontologist to confirm and enhance his identifications.

You may also find vertebrate fossils inland that are accompanied by shells. You may be working where there was once an ocean, beach, marsh, or perhaps a stream or river. Many shells that you find on the same spoil banks or flats as you do your vertebrate fossils are of great interest to the hunter of invertebrates. Pick some up. Even though you know little about them, you can certainly give them away and you will soon learn what you find often or what is rare — at least in your vicinity.

Shells are also a concern of the vertebrate paleontologist. He knows whether they are fresh, or salt water species. In conjunction with the vertebrates found with them, he can tell whether the animals lived and died in or near salt or fresh water. Shells can be dated as well as vertebrates. The pros are interested when you find comparatively young vertebrate fossils in the same locale as very old shells. The reverse is equally true.

CHAPTER V

The "rules of the game" are left until the end so that you will remember them.

1. Always take at least a gallon of water with you. Not only will you need it but the radiator may, too. Don't have extra walking forced on you by a leaky radiator. These things develop at strange and inopportune times. Be certain that the spare is OK and that you have tire changing equipment.

2. Always wear rugged shoes. The sedimentary deposits, of which fossil spoil banks consist, may become very hard as they weather out. It will be necessary to climb to the top of many. These are not gentle slopes, neither are they soft sands and banks. They can be very slippery.

3. Always wear a hat. In the summer time, spoil banks may cut off the breezes and the sun bakes down where you are working. Make sure that the hat has a brim all the way around it. You will be bending over so no billed fishing or hunting cap will protect the back of your head. You will find that you must shade your eyes with a hat on cold days, too. If the sun is bright, so is the glare on the spoils. No matter the cost of dark glasses, when you wear them fossils seem to blend into their background and are far more difficult to distinguish. Hunt as much as possible during the times when there are few or no shadows on the fossils, not too early or too late in the day.

4. Watch out for snakes in snake country. Don't pull yourself to the top of a pile until you can see that the ground is clear where your hand will touch. Snakes sun-bathe and like water. Watch for them at the water's edge beside the banks. Flats have the added advantage of few snakes and those you can see a long way off.

5. Always carry a light-weight, long handled, garden tool that has some sort of curved tines or scoop on the working end of it. It will have many uses. You can "pick" out completely exposed fossils that are out of reach. It can be a third leg to help with balancing on a hog-backed mound. It can go through high grass or brush ahead of you. Probably its most useful function is tapping, GENTLY, what appears to be a fossil. More than half the time it will prove to be a leaf, piece of bark, shell or stick. Only a fossil "sounds" like a rock when you tap it. A shell sounds like tapping on china. The other things just "thunk". Save yourself from unnecessary bending. There will be more than enough of that when it's necessary to pick up the real thing.

6. Always carry a bucket or knapsack. Pockets are neither safe nor large enough. Take paper in which to wrap fragile specimens. If you plan to hunt in hard deposits, take a small unbreakable pick such as is used by a rock hound. A trowel is a help in softer material. NEVER, NEVER take hold of partially exposed fossils and PULL. Some will break like glass. If you chip with your pick or dig with your trowel three or four inches away from small fossil bones or teeth and farther from large ones, they will come loose in their matrix and can be retrieved undamaged from it. Even the concussion that radiates from any blow too near fossils can

cause the specimens to shatter. This is an avoidable tragedy. Take your time. Fossils will wait.

7. Take somebody with you, stay within calling distance and preferably in sight of your companion. However, as on the beach, stay far enough apart to discourage chatting or constant interruption. Both are blinders and upset the concentration that is vital to what you are doing.

As in most sports and outdoor hobbies there are things that can go wrong. You can slip or trip. Shells cut. Use your third leg to be certain a water's edge is firm before you walk on it.

Be ready to help and stay close enough so that somebody will know if you are in trouble.

8. Catalogue your fossils. Print a number with a contrasting colored indelible ink on each fossil and the corresponding number in a notebook. Next to it list its identification, the location in which it was found, and when you found it. No paleontologist is interested in any uncatalogued collection. The particulars of each specimen are as important to him as is the packaging that you expect in your grocery store. Consider your feelings if everything there was just in piles—no descriptions. Cataloguing is "packaging" to a pro. It, too, tells the whole story of the "product".

9. There is little use in hunting newly dug spoil banks. Seldom do fossils or shells show through until there has been plenty of wind and rain. This is not so of fossils found on flats. When they have come through the dredge spoil pipes they can be seen as soon as they hit the ground. Later they will be covered by the accompanying mud and then you will have to wait until they, also, weather out.

10. Many of your fossils will need to be preserved. They are old and now that they are exposed to the air they may begin to crumble. There are several ways in which to treat them. One of the most simple is to cut the thickness of Duco cement with acetone and paint the thinned (but not watery) mixture all over the fossil. Be certain that the fossil is dry before you do this. The acetone will help to "pull" the glue *into* the fossil. Work outdoors with acetone. It is dangerous to inhale and extremely inflammable. Do not cover your fossils with shellac. There is no use in preserving only the outside of the fossil. If the inside rots away there will soon be no outside, anyway. When properly applied, the aforementioned mixture will also preserve the original looks of the specimen.

You are starting on a hobby that can be hot, cold, dirty, buggy and wet as you clamber, crawl, slide and stumble up and down the banks.

You are also starting on a hobby that provides plenty of exercise, new friends who are good sports, complete oblivion to outside worries, ever increasing knowledge and wonder concerning a fascinating earth science, and FUN.

However, there will be little fun in watching helplessly as the banks, that you know are still loaded with fossils, are hauled away to be used for foundations for new roads and bridges.

It will be equally distressing to see a flat that is still producing fine specimens after each rain, slowly but surely become a well planted and then occupied trailer park. Don't waste time. When grass seed is sown on a flat you will have to leave it alone.

Remember your motto:

"Business is picking up!"

So — start walking.

List here any fossiliferous beaches or banks that you find and that are not described at the end of the book:

COASTAL STATES OF THE U. S. WITH A CURRENT REPORT ON THEIR BEACHES

MAINE

The fossiliferous beach areas have not been mapped so cannot be listed individually. However, along the Kennebec River a number of sites have been located where Pleistocene fossils are found. These reach from Norridgewock southward to the present coastline.

To the best of the knowledge of marine research experts in Maine, the only vertebrate fossils found there have been those of the walrus—and only a very few of those.

If you are interested, get the reports from the Woods Hole Oceanographic Institute concerning fossils dredged from the bottom of the Gulf of Maine.

Maine has invertebrate fossils that are older than the Pleistocene. (Report any beach or bank fossil sites to: Dept. of Sea and Shore Fisheries, State House, Augusta, Maine.)

MASSACHUSETTS AND RHODE ISLAND

Anyone in this territory will be able to hunt in these locations simultaneously, so they are listed together. Ancient bones and invertebrates may be found at Martha's Vineyard off Cape Cod, the Elizabeth Islands, Nantucket, No Man's Land, and Block Island. A boat will be handy to get to some of these. (Report new sites to: Rhode Island Development Council, State House, Providence 2, Rhode Island.)

CONNECTICUT

Fossils on beaches here so far unknown. The rocks that are adjacent to the beaches of Connecticut are so highly metamorphosed that the fossils in them have probably been destroyed. Thus, there is very little chance of finding any. (Report any beach fossil sites to: State of Connecticut, Geological and Natural History Survey, Box 128 Wesleyan Station, Middletown, Connecticut 06457.)

NEW JERSEY

Atlantic Highlands in Monmouth County is good. The bluffs along Sandy Hook Bay contain invertebrates. Also, sharks' teeth, spinal discs and various coprolites are not uncommon. Occasionally fish teeth and vertebrae and turtle bones are found. If you are really lucky you may find bones of much older animals than are described in this book.

Shark River is not quite on the coast but is close and has fossiliferous spoil banks and river banks. Sharks' teeth and the remains of ancient fish are hunted here.

The oldest fossil ants yet discovered were preserved in amber and found by amateur hunters in the Cliffwood Beach area of Monmouth County. If any of you find any fossil insects and/or amber, please report your find and the site to: Dr. Frank M. Carpenter, The Biological Laboratories, 16 Divinity Avenue, Cambridge, Massachusetts 02138.

Be careful on New Jersey beaches not to mistake recent remains of leached or slightly altered vertebrates or invertbrates for true fossil forms. Here it is a tricky and trying business. Get a pro's identification and confirmation. (Report any new beach, bluff, or bank fossil sites to: Museum of Natural History, Geology Dept., Guyot Hall, Princeton University, Princeton, New Jersey 08540.)

NEW YORK

The geological survey reports that they have made no beach survey. Try some of the Long Island beaches that have bluffs near them. Check the bluffs, particularly. (Report any beach or bluff fossil sites to: Geological Survey, New York State Museum and Science Service, State Education Building Annex, Albany, New York 12224.)

DELAWARE

Along Delaware's Atlantic shore (25 miles or so between Cape Henlopen and Fenwick Island) the beach materials are largely derived from deposits of Pleistocene age by reworking. Fossil trees are exposed along portions of the beach under certain storm conditions and it is possible that some Pleistocene shell materials are present along the recent beach; however, they know of no fossil vertebrate remains ever being recovered from the Delaware beaches. (Report any other beach fossil sites to: Delaware Geological Survey, University of Delaware, Newark, Delaware 19711).

MARYLAND

This is the state that has the world famous Calvert and Scientist Cliffs. Vertebrate fossils may wash out of them onto the beaches, or if they have already done so, may have been swept out into Chesapeake Bay and be washed in again for you to find along the edges or in the shallow water. The cliffs contain many vertebrates, and even more invertebrates, of Miocene age (twenty million years). However, the Pliocene (an epoch older than the Pleistocene but younger than the Miocene) and the Pleistocene show at the top of the cliffs so you may find the remains of three ages at their bases.

For specific information send for Bulletin Number 28 from the Dept. of Geology, Mines and Water, Johns Hopkins University, Baltimore, Maryland. This has pictures, too.

Report any exceptional finds to them.

It is strictly forbidden for amateur collectors to dig in the actual cliffs. Don't do it as you will be trespassing on professional territory. Also, do not cross privately owned land to get to the beaches. There are public entrances.

No matter how fossiliferous the cliffs are now, amateur collectors who remove fossils from them for their own collections may be destroying a chapter of yet unrecorded or documented history that is being studied by many professional geologists and paleontologists. You are in their classroom. Unless you have specific and written permission from a Maryland institution of higher learning and from the Smithsonian Institution, Washington, D. C., LEAVE THE CLIFFS TO THE PROFESSIONALS.

VIRGINIA

Ocean beaches produce few fossils in Virginia. However, fossils millions of years older than those of the Pleistocene epoch (Miocene) can be found along the James River in the vicinity of Williamsburg and along the shores of the York River in the vicinity of Yorktown. Much of this land is privately owned so be sure to get permission before you start to hunt. A collecting site known as the "Kenneth E. Rice Memorial Museum and Fossil Pit", located near Hampton, Virginia, has been opened (for a small charge) to the public. Write to them for a brochure. Also, check locations with the Richmond, Virginia Gem and Mineral Society, Inc. (Report any other beach or bank fossil sites to: Division of Mineral Resources, Natural Resources Building, Box 3667, Charlottesville, Va. 22903.)

NORTH CAROLINA

The Coastal Plain of North Carolina has abundant fossil forms. The eastward sloping formations extend out under the ocean. Current and wave action serve to loosen fossils and pitch them on the beach.

From experience it appears that the bay and inlet areas offer better collecting sites. (Report specific beach and bank fossil sites to: Museum of Natural History, Raleigh, N. C. 27602.)

SOUTH CAROLINA

There are fossiliferous beach deposits from the Myrtle Beach area southward. The state's greatest limestone deposits are located in the Orangeburg vicinity not too far from the coast, if you want to try that, too. Fine collections have been gathered from the south end of Pawley's Island and from Litchfield Beach just above Pawley's. Edisto Beach which is south of Charleston is an excellent fossil bearing beach of Pleistocene epoch material. (Report other beach and bank fossil sites to: The Charleston Museum, Charleston, South Carolina 02813.)

GEORGIA

There are fossils near coastal beaches in Georgia. Jekyll Island and banks of the inland waterway at the Isle of Hope are good. Also check out the Kingsland - St. Mary's area and Whiteoak in Camden County. Stokes Ferry 11 miles south of St. George, reports fossils. There are several localities in barrow pits along interstate #95—particularly at Savannah. These may be on private land so be certain to get permission to hunt there. (Report any beach or bank fossil sites to: The Dept. of Geology, University of Georgia, Athens, Georgia 30601.)

FLORIDA

For coastal beach and bank fossil hunting, this state offers more than any other to the amateur collector.

Fossils may be found on the spoil banks of the St. Johns River, Jacksonville. They are also hunted on the coastal beaches from the Georgia state line down past Vero Beach on the east coast and near any inland canals or on spoil banks fairly close to the sea even farther south.

On the west coast there are too many vertebrate fossil sites to list. Both beaches and locations near beaches from New Port Richey, through St. Petersburg, Tampa, Bradenton and Sarasota are worthwhile studying. Venice, in Sarasota County, provides the best beach fossiling in the entire state. All canal and spoil banks in this area are excellent sites, also.

As you drive farther south the fossiliferous locations are more numerous inland in modern diggings than they are on the beaches. Look for new canals and "flats" close to Highway #41. There you may find both vertebrate and invertebrate fossils. Don't invade fenced land without permission from the owner. In Florida there is a great deal of fenced grazing land for cattle, Neither the owners nor the cattle want trespassers. The majority of the grazers are not friendly hosts. Stay out! (Report other specific beach or bank fossil sites to: The Dept. of Natural Sciences, Florida State Museum, Seagle Building, Gainesville, Fla. 32601.)

ALABAMA

Vertebrate fossils so far not found on Alabama beaches. However, some invertebrates are scattered thinly. See what you can find. (Report new sites to: Geological Survey of Alabama, University of Alabama, University, Alabama 35486.)

MISSISSIPPI

No vertebrate fossils yet reported from the beaches which are very low and formed about six thousand years ago. (Report any beach or bank fossil sites to: Gulf Coast Research Laboratory, Ocean Springs, Mississippi 39564.)

LOUISIANA

Two areas here provide fine invertebrates. They are easily accessible and the supply of fossilized shells, casts and molds is abundant. The localities are at Grand Isle, off the coast of Jefferson Parish, and along the coast of Cameron Parish in southwest Louisiana. No vertebrates yet reported. (Report new sites to: Louisiana Geological Survey, Louisiana State University, Baton Rouge, Louisiana 70800.)

TEXAS

Almost anywhere that there is active erosion along the coast can be a good place to hunt vertebrate fossils. Look particularly near the mouths of the rivers that empty into the Gulf of Mexico. Both the University of Texas and the Geological Survey say that the overall picture is good but no specific sites are yet listed. (Report beach and bank fossil sites to: The Dept. of Geological Sciences, the University of Texas at Austin, Austin, Texas 78712.)

CALIFORNIA

Ask the cost first, then prepay a request for Chapter III, "Historical Geology", of Bulletin 170, "Geology of Southern California". Address: Division of Mines and Geology, Ferry Building, San Francisco, California 94100.

California fossils and locations are discussed in this, and the fossil deposits in California are too extensive to cover here. Most fossil beach areas are noted for marine and non-marine invertebrates, but there are three near-coastal areas in southern California where bones have been found. These are the La Brea Tar Pits, of course, and Carpinteria, as well as the coastal area of Santa Rosa Island. You can't hunt at LaBrea but you should certainly see their museum. There are vertebrates found at Newport Bay in Orange County, and at San Pedro, Playa del Rey, and Santa Monica (Los Angeles County). (Report beach and bank fossil sites to: Los Angeles County Museum, 900 Exposition Blvd., Los Angeles, California 90007.

OREGON

The best place to look for fossils along the beaches of Oregon is in Lincoln County, particularly at Beverly Beach State Park and in the sea cliffs for the next three miles south of the park area. Fossil sea shells of Miocene age (20 million years old) are abundant here, and the bones of whales and sea lions are sometimes found inside hard concretions. Fogarty Creek State Park a few miles north has similar fossils in its sea cliffs. Fossil shells of Oligocene age (30 million years old) are hard to find but fun to look for on Short Sands Beach at Oswald West State Park in Tillamook County. Many fossil shells and occasional shark teeth of Eocene age (40 million years old) can be collected at Sunset Bay and Cape Arago State Parks in Coos County. The jaw and teeth of an ichthyosaur, a swimming cousin of the dinosaur, were found recently in Jurassic rocks (150 million years old) south of Humbug Mountain State Park in Curry County. Rare discoveries such as this are of great scientific value. Once in a while a beachcomber finds a tooth of an Ice-Age elephant exhumed by storm tides from an ancient coastal peat bog. All of these locations are shown on state road maps. (Report other beach and bank fossil sites to: Dept. of Geology and Mineral Industries, 1069 State Office Building, Portland, Oregon 97201.)

WASHINGTON

It will be a rare find if you discover vertebrate remains on Washington beaches or near there. One of the most interesting collecting areas is found along the sea cliffs west of West Twin Rivers in Clallam County. Here, one can find some vertebrate material. It consists of agatized or permineralized bone fragments and masses embedded in the Oligocene marine sediments. They probably come from a marine animal but no identification has been made. Several miles to the west, near Slip Point on Clallam Bay, a cranium of a dolphin-like animal was found in 1957 along with cetacean bonês. This was in the Miocene Clallam Formation. Tusks and teeth of mammoth and mastodon have been found in Pleistocene sediments along the north side of the Olympic Peninsula at Sequim Bay and teeth on the beaches on the west side of Whidby Island in Puget Sound. (Report any other beach or cliff fossil sites to Dept. of Natural Resources, Box 168, Olympia, Washington 98501.

ALASKA

Most fossil localities are remote areas due to the primitive state of the communications network. A compilation of roadside or beach localities has never been attempted in the state and, therefore, there is no accessible information of this kind.

Peruse the various U. S. Geological Survey bulletins and professional papers which deal with the paleontology of Alaska if you wish to pinpoint some of the more important localities, regardless of accessability. The publications could be viewed at the large universities in your state. (Report all beach and inland fossil sites to: College of Earth Sciences and Mineral Industry, Dept. of Geology, University of Alaska, College, Alaska 99701.)

HAWAII

There are no vertebrate fossils on beaches in Hawaii. This is explained by the volcanic deposits that are so prevalent there. However, there are invertebrates that can be found on some of the beaches. (Report exceptional beach or bank fossil finds to: University of Hawaii, Dept. of Geology, Honolulu, Hawaii 96800.)

RECOMMENDED READING

Now you have had an introduction to a hobby and the barest glimpse of the science of vertebrate paleontology. Don't stop here. Increase your enjoyment and knowledge with the following books. There are many, many more—too many to list. These will help YOU—and that's what really counts.

THE EVIDENCE OF EVOLUTION: by Nicholas Hotton III, published by the American Heritage Publishing Co., Inc., New York, New York, in association with the Smithsonian Institution, Washington, D. C.

Get this first!

155 pages of stimulating reading, fascinating information and magnificent colored illustrations of fossils and living specimens. One of America's leading paleontologists, a curator at the Smithsonian Institution, writes for the intelligent layman. He tells it as it was and as it is with the excitement and enthusiasm that only a great scientist and writer can share so completely with an amateur. The price is most reasonable.

FOSSILS: An Introduction to Prehistoric Life, by William H. Matthews III, Published by Barnes & Noble, Inc., New York.

This is a book that all amateur fossil collectors must read. It is paperback. It is inexpensive and you can understand it and enjoy it. Written by a scientist who is a friend of amateurs, it is accurate, thorough and interesting. Well illustrated and with most helpful lists of pertinent publications, localities, collections, etc.

HORSES: by George Gaylord Simpson, published by Doubleday and Company, Inc., Garden City, New York.

This book can be bought in hardback or in paperback. It explains to amateur (as well as to professional) collectors everything that is pertinent to the horse family; its history and how it has changed, progressed and developed for millions of years right up to the present. Well illustrated and written as only one of the scientifically crowned "greats" of vertebrate paleontology can. This book benefits every inquiring amateur who wants to learn more about the fossil bones and teeth of the ancient horses that roamed so many parts of the United States for so long. Only this expert could make it all so clear and easy to understand.

THE FOSSIL BOOK: by Carol Lane Fenton and Mildred Adams Fenton, published by Doubleday & Co., Garden City, New York.

This is a big book and expensive. It is profusely illustrated with drawings and photographs of a selection of both vertebrate and invertebrate fossils. This is a good companion on any fossil collecting trip across the United States as it touches on all Ages, many sites and much pertinent extracurricular reading material. As it covers so much ground it does not specialize in any particular animal.

BONES FOR THE ARCHAEOLOGIST: by I. W. Cornwall, published by Phoenix House, Ltd., London, England.

This is the book that actually teaches you how to identify fossil bones and teeth of a good cross section of some vertebrate families. The illustrations are so superior that you can see each bone, separately, though still in place in the complete articulation of various parts of their entire skeletons. You are now ready and able to understand the text though the abbreviations and measurements in meters may slow you up. This book is not inexpensive in cost but the value of what you can learn from it far exceeds the price. It's truly one of a kind and a MUST.

VERTEBRATE PALEONTOLOGY: by Alfred S. Romer, published by the University of Chicago Press, Chicago, Illinois.

This is the backbone of vertebrate paleontology. A new, revised and well illustrated edition is now on the market. This book is used as a text in most first class colleges in which the subject is taught. There is no other that is so indispensable in any pertinent collection of v.p. books. Here, you can find answers to your toughest questions. The book is written by one of the most learned, respected and experienced men in the field. He is a teacher of the pros as well as of the potential pros. You must have a foundation of learning from the other books listed before you can truly appreciate "Vertebrate Paleontology". However, as a reference, if you can read an index you need it right NOW.